The University of Tokyo Studies on Asia

The University of Tokyo Studies on Asia series welcomes submissions from scholars in Asian Studies from across the world. We publish research monographs and edited volumes that offer comprehensive, high-quality scholarship on Asian art, politics, societies, history, and cultures, with particular emphasis in literature; social and cultural history; anthropology; visual studies; law; philosophy; and religious studies. This series is unique for its inclusion of North, East, South, Southeast and West Asia (also commonly known as the Middle East) in its remit. With an emphasis on Asian voices and moving beyond the traditional dominion of Asian Studies, this hybrid series publishes both Open Access and regular market books with particular emphasis on the utilitarian value of developing pathways for authors who would not normally consider Open Access. While we will offer a market book option, we are dedicated to the OA principles through which research outputs will be distributed online, free of cost, to readers across the world without barriers to copying or reuse. We aim to make Asian Studies research by Asian freely available for anyone who might benefit.

We aim to solicit works from authors and editors whose studies focus on individual Asian countries as well as regional and trans-regional topics. We define the boundaries of Asian Studies quite broadly – from North to South and East to West. The Series Editors curate a theoretically rich series deeply engaged with postcolonial theory, gender studies, and cultural studies, led by the expertise presented by the scholars, and affiliated network, of the series' home: the Institute for Advanced Studies on Asia. To date, postcolonial theory and gender studies have been prominent in scholarship by Anglo-phone scholars in the US, UK, and India. We aim this series to promote opportunities for new theoretical work by scholars from Korea, China, Southeast Asia, and Pacific-Asia.

The series editors invite submissions of original works across the broad spectrum of humanities and social sciences as well as translations of established works not previously published in English. Based at the University of Tokyo Institute for Advanced Studies on Asia, the series seeks to engage scholarship 'on Asia in Asia', with the intent of bringing forth the intellectual works of those not otherwise likely to be known to scholars whose primary language is English.

We welcome contributions to cutting-edge issues and debates from younger researchers as well as from established scholars. We are particularly interested in submissions that cross disciplinary boundaries or seek to establish new disciplinary methods.

This book is supported by funds from the UTokyo Foundation, and we would like to express our sincere thanks to the donors.

Xue Qu · Daizo Kojima · Laping Wu ·
Mitsuyoshi Ando

Harvest Loss in China

Rice, Mechanization, and the Moral Hazard of Outsourcing

Springer

Xue Qu
Xi'an Jiaotong University
Xi'an, China

Daizo Kojima
The University of Tokyo
Tokyo, Japan

Laping Wu
China Agricultural University
Beijing, China

Mitsuyoshi Ando
The University of Tokyo
Tokyo, Japan

ISSN 2731-7633 ISSN 2731-7641 (electronic)
The University of Tokyo Studies on Asia
ISBN 978-981-97-9155-2 ISBN 978-981-97-9156-9 (eBook)
https://doi.org/10.1007/978-981-97-9156-9

This Springer imprint is published by the registered company Springer Nature Singapore Pte Ltd.
The registered company address is: 152 Beach Road, #21-01/04 Gateway East, Singapore 189721, Singapore

Acknowledgments

This book is the result of a joint research by the University of Tokyo and China Agricultural University. Some chapters are based on the authors' previously published articles (in journals Agriculture, Agricultural Economics - Zemedelska Eknonmika, Applied Economics, International Food and Agribusiness Management Review, Japanese Journal of Farm Management, Journal of Integrative Agriculture, and Sustainability), which have been substantially rearranged and revised for this book. We must first thank our colleagues in the Department of Agricultural and Resource Economics, the University of Tokyo, specifically Profs. Akira Kiminami, Takenori Matsumoto, Katsuhiro Saito, Yasuhiro Nakashima, and Takeshi Sakurai, Associate Professors Hironori Yagi, Kentaro Kawasaki, Tomoaki Nakatani, and Takao Yurugi, and Assistant Professor Yukinaga Nishihara for their highly valuable and detailed suggestions that helped enhance this work.

Additionally, we express our gratitude for the generous funding support received from a variety of sources. The 2015 special scientific research project of grain public welfare industry—"Investigation and evaluation of rice harvest loss" (No. 201513004-2)—supported the critical preliminary data survey for this work. More than 1000 people participated in the survey, including project team members, recruited students, and villagers. We express our heartfelt thanks to all those who helped and guided us in this project, specifically our colleagues in the College of Economics and Management, China Agricultural University, Profs. Junfeng Zhu, Jun Li, Ji Ma, Hongman Liu, and Hailong Cai, and Associate Professor Zhiwang Lv. Subsequent work was also supported by JSPS KAKENHI (Grant Number JP19H03063) and a scholarship from the China Scholarship Council.

Our thanks also go to the book series team of the University of Tokyo Studies on Asia at Springer Nature Press: Series Editor Yasuhiro Matsuda and Executive Editor Christopher Gerteis, who initially guided us through the preparation of our book submission and later supported the book publication. Senior Editor Alex Westcott Campbell and Production Editor Bharath Kumar Dhamodharan have been supportive and effective editors and greatly helped by guiding us through the publishing process.

Indeed, the assistance received for this book over the five years since its inception far exceeds that mentioned above. We express our gratitude to all these individuals and organizations for their assistance.

June 2024

Xue Qu
Daizo Kojima
Laping Wu
Mitsuyoshi Ando

Contents

Abbreviations

2SLS	Two-Stage Least Squares
CFS	Committee on World Food Security
FAO	Food and Agriculture Organization of the United Nations
MARA	Ministry of Agriculture and Rural Affairs of China
MCMC	Markov Chain Monte Carlo
NBSC	National Bureau of Statistics of China
OLS	Ordinary Least Squares
Planning	Regional Layout Planning for Advantageous Agricultural Products
PSM	Propensity Score Matching
RCRE	Research Centre for Rural Economy
RFOP	Rural Fixed Observation Point
U.S.EPA	The United States Environmental Protection Agency
USDA	The United States Department of Agriculture
WRI	World Resources Institute

List of Figures

List of Tables

Chapter 1
Introduction

1.1 Background and Significance

1.1.1 Food Wastage and Food Security

With a large and growing population, ensuring an adequate food supply has always been one of the most important goals worldwide. According to the Food and Agriculture Organization of the United Nations (FAO), food production will have to increase by 70% to feed the world's projected 9.1 billion population by 2050 (FAO 2009). In 2018, this world population projected figure has been updated to 10 billion by the World Resources Institute (WRI) (Ranganathan et al. 2018). As the most populous developing country in the world, China's food security is of great concern to the world.

There are two ways to increase food supply. For a long time, the dominant view for improving future food supply has been to increase food production; however, this would not achieve the world's growing agricultural demand in an environmentally sustainable manner (Shafiee-Jood and Cai 2016). Agriculture imposes huge resource and environmental costs in terms of land (Chen et al. 2014), water (Pierrat et al. 2023), and greenhouse gas emissions (Xu et al. 2021). The trade-off between food security and environmental sustainability will aggravate in the near future with the diets change (especially meat consumption), bioenergy crop expansion, as well as climate change (Godfray et al. 2010; Foley et al. 2011). Therefore, improving food supply by increasing food production is limited by the progressively scarce natural resources and the fragile environmental capacity (Qu et al. 2021).

The other way to increase food supply is to reduce food wastage (food loss and food waste) (Hodges et al. 2010; Barrera and Hertel 2021), which is too large to be ignored. A report by FAO has pointed out that globally, about 1.3 billion tons of food for human consumption are lost or wasted each year, accounting for one-third of the world's food production and enough to feed 3.48 billion people (Gustavsson et al. 2011). After the food crisis of the early 1970s, preventing food wastage has been

X. Qu et al., *Harvest Loss in China*, The University of Tokyo Studies on Asia,
https://doi.org/10.1007/978-981-97-9156-9_1

widely recognized as a solution to the world's food problems (FAO 1981; Greeley 1991). In 2015, FAO proposed Sustainable Development Goal 12.3, "By 2030, halve per capita global food waste at the retail and consumer levels and reduce food losses along production and supply chains, including post-harvest losses" (FAO 2015). In 2019, the United Nations designated September 29 as the International Day of Awareness of Food Loss and Waste. In 2021, Chinese government issued the *Anti-Food Waste Law of the People's Republic of China*, placing a high priority on food wastage reduction.

Although the reduction of food wastage to meet food supply appears to be a second option that is constrained by resources and the environment, reducing food wastage can, in turn, be beneficial to resource and environmental sustainability. In other words, food wastage implies waste of resources and damage to the environment. Food production, processing, and transportation consume enormous quantities of resources. If these lost and wasted foods are not eventually consumed, the resources embedded in the food are wasted, which contributes to environmental degradation (Bai and Dent 2006; Jalava et al. 2016; Shafiee-Jood and Cai 2016; Yu and Jaenicke 2020). It is estimated that nearly a quarter of the water, arable land, and fertilizer used globally for food production is associated with food wastage (Kummu et al. 2012). In 2010, China's food waste accounted for more than 10% of the country's total water use (Sun et al. 2018b). The wastage in the entire supply chain of grain in China means about 26 million hectares of land were used in vain, which is equal to Mexico's total arable land (Liu et al. 2013). With global food wastage causing 3.3 Gtonnes of carbon dioxide equivalent emissions, if regarded as a country, food wastage would be the third-largest emitter after the United States and China (FAO 2013). Therefore, reducing food wastage can achieve increased food availability without investing additional resources, such as land and water, thus reducing the harm to the resources and the environment.

1.1.2 Importance of Rice Production

Rice, as one of the world's three most important food crops, feeds almost half of the world's population, making it the most consumed cereal grain (Arvanitoyannis and Tserkezou 2008; Qian et al. 2023), particularly in developing countries (Maclean et al. 2013). China is the world's largest producer of rice, with 212 million tons produced in 2020, accounting for 30% of the total world production (FAOSTAT). In China, rice has received a much higher priority in government policy, which is required to be "absolutely safe" given its status as a staple food.

In developing countries, more food is lost at the front end of the supply chain (Gustavsson et al. 2011). Along the food supply chain, farmers suffer more losses than processors and marketers (Bala et al. 2010; Babatunde et al. 2019; Delgado et al. 2021). Such losses suffered by rice farmers reduce their incomes and threaten their food security, especially for farmers in developing countries living on the edge of hunger (Coker and Ninalowo 2016; Danbaba et al. 2019). Among the losses occurring on farms, a higher proportion of rice is lost during harvesting than during other stages on the farm (Basappa et al. 2007; Lu et al. 2022). Most farmers lose a significant amount of rice during harvesting (Ibrahim et al. 2018). As the beginning step of the grain supply chain (Kantor et al. 1997), harvesting is the decisive step that determines the quality of the yield (Sawicka 2019; Müller et al. 2022) and the subsequent storage (Tefera 2012). However, more than 80% of known research focus solely on on-farm storage losses (Affognon et al. 2015), implying that losses during harvesting process have received less attention. Thus, this book focuses on rice harvest losses.

1.1.3 Booming Machinery Outsourcing Service

Agricultural production in China is a typical smallholder economy (Ren et al. 2023) with an average farm size of 0.5 hectares (Zhang et al. 2017), which is considered unfavorable for mechanization (Ruttan 2000; Pingali 2007). However, since the reform and opening-up in 1978, mechanization has grown rapidly in rural China, especially in the last two decades. As Table 1.1 gives, from 1980 to 2019, tractors increased from 2.61 million to 22.2 million at an annual growth rate of 5.64%. The number of combines has increased 71 times from 0.03 million to 2.13 million.

The rapid development of agricultural mechanization in China benefits from the machinery outsourcing services (Sheng et al. 2017; Wei and Lu 2024). Over the past few decades, China's urbanization has led to a massive exodus of the predominantly young male rural labor force, resulting in an aging and feminized rural labor force (Rozelle et al. 1999; Xie and Lu 2017; Xu et al. 2019). Rapid declines in fertility and mortality have exacerbated the aging of the rural labor force (Li and Sicular 2013). The decline in both the quantity and quality of rural labor force accompanied by aging and feminization has led to an unprecedented demand for machinery operation by farmers (Tan et al. 2019), especially in labor-intensive stages (Deng et al. 2020), such as land preparation and harvesting. In addition, the accelerating rise in the relative price of rural labor against machinery has also created the conditions for the booming development of machinery outsourcing services in China (Wang et al. 2016; Yi 2018). Although the average farm size is small, the specialized and divisible nature of agricultural production stages allows farmers to shift some production stages or hire others to undertake some production tasks (Mi et al. 2020), which allows some specific production stages to be carried out on a larger scale, such as harvesting (Zhang et al. 2017). As a result, an outsourcing services that rent out the package of

Table 1.1 Mechanization in China (million)

	1980	2000	2010	2015	2019
Tractors	2.61	13.62	21.78	23.10	22.2
Combine harvesters	0.03	0.26	0.99	1.74	2.13

Source National Bureau of Statistics of China

agricultural machinery and machine operators came into being (Belton et al. 2018; Cai and Wang 2021). Machinery outsourcing services allow farms of all sizes have access to agricultural machinery (Belton et al. 2018; Wei and Lu 2024), helping small farms overcome their size disadvantage (Picazo-Tadeo and Reig-Martínez 2006).

Therefore, although small-scale farmers may not be able to own expensive agricultural machinery, they can afford relatively cheap harvest outsourcing services. From 2003 to 2005, the area harvested by cross-regional agricultural machinery services grew from 6,794.69 khm^2 to 16,522.92 khm^2 (Deng et al. 2020). In 2017, more than 280 million hectares of farmland were served by harvest outsourcing service organizations or individuals (MARA, 2018). In 2020, there were 194,600 agricultural machinery service organizations, including 78,900 agricultural machinery cooperatives; 40.08 million farm households engaged in providing the agricultural machinery services, including 4.23 million agricultural machinery specialized service households. Specifically for rice, a national study showed that about 70% of farmers outsourced their harvesting operations (Wang et al. 2011). In Anhui, this percentage was close to 85% (Cai and Cai 2014).

1.2 Problem Statement and Research Objectives

When studying harvest losses, machinery is a factor that cannot be ignored. Machines operated by humans are easier to control compared to other factors, such as weather and pests. However, the effect of machine on harvest losses has not been conclusive. Gao et al. (2016) thought that machines help reduce rice harvest losses. Greeley (1982) stated that the increase in threshing losses of rice in Bangladesh was related to technological change. Both the results of field surveys (Li et al. 1991; Huang et al. 2018) and farmer survey (Zhan 1995) in China showed that the rice harvest losses of mechanical operations were greater than manual operations. A common feature of these studies is that they all ignored the harvest outsourcing services behind mechanical harvesting, which may have a negative effect on harvest losses.

Mechanization achieved by outsourcing services is not only a substitution of machine for labor, but also a substitution of market labor for household labor (Zhang et al. 2017; Yi 2018; Cai and Wang 2021). Through the purchase of harvest outsourcing services, a principal–agent relationship is formed between farmers and service providers, in which farmers are principals and service providers are agents (Huan and Hou 2020). Moral hazard is prevalent in outsourcing services in the secondary and tertiary industries (Andersson et al. 2019; Lu and Du 2020). Therefore, there is no reason to suppose that outsourcing services in agricultural production do not have moral hazard problem (Lu and Du 2020; Yue et al. 2023).

First, the aims of farmers and service providers are not aligned. Farmers want to harvest all their crops quickly and economically. However, the goal of service providers is to maximize their own profits, which is based on the harvested area (Zhang et al. 2017; Li et al. 2023), regardless of factors such as crop yield and quality, decline in land fertility, and pollution (Huan and Hou 2020). As a result, service providers may perform harvesting work in a rough manner, e.g., by increasing the harvesting speed of the machine, ignoring rice in the corners, and neglecting to adjust the machine. Second, due to the unregulated outsourcing service market and the general lack of written contracts (Cai and Liu 2019), farmers do not have complete information about service providers' skills and efforts or the maintenance status of machine, resulting in information asymmetry (Yang 2007; Sun et al. 2018a). Further, it is difficult for farmers to monitor the service quality of service providers (Tocco et al. 2012; Wang et al. 2022). Although farmers can observe harvest losses after the harvesting, retrospective observations do not provide effective monitoring. Therefore, as hired workers, service providers do not take their work as seriously as farmers (Coelli and Battese 1996; Sun 2013). They may adopt labor-saving and sloppy practices (Cai and Liu 2019).

Despite the increasing competition in the outsourcing service market due to the increase in the number of machines and the cross-region harvesting, China's harvest outsourcing services remain a seller's market (Shen et al. 2015). Farmers are still at a disadvantage and disputes regarding service quality are frequent (Han 2019). Some scholars provided the evidence of moral hazards of service providers from the perspective of game theory (Cai and Liu 2019; Huan and Hou 2020). Chegere (2017) also indicated that the conscientious attitude of machinery operators would affect the magnitude of harvest losses. However, there is a lack of empirical studies on moral hazards in harvest outsourcing services. With the increasing prevalence of harvest outsourcing services, it is necessary to study the moral hazard involved and its effect on rice harvest losses.

In summary, this book aims to estimate rice harvest losses in China and study the moral hazard in harvest outsourcing services and its effect on rice harvest losses. Specifically, the objectives of this book are as follows:

Objective 1: To estimate the magnitude of rice harvest losses.

Objective 2: To study the moral hazards in harvest outsourcing services.

Objective 3: To explore the effect of harvest outsourcing services on rice harvest losses and investigate whether the effect of harvest outsourcing services on rice harvest losses is achieved through moral hazards.

Objective 4: To analyze the effect of moral hazards in harvest outsourcing services on rice harvest losses.

1.3 Related Concepts in Context

1.3.1 Harvest Loss

Before discussing harvest losses, it is necessary to make a clear distinction between food loss and food waste. In fact, the food wastage mentioned at the beginning of the book includes food loss and food waste (FAO 2013; Timmermans et al. 2014). Some studies also used the singular term "food loss" or the terminology "food loss and waste" without distinguishing between food loss and waste (Minor et al. 2020).

Food loss and food waste are two different concepts that occur for different reasons and at different locations along the supply chain. Food waste occurs at the back end of the supply chain, primarily at the consumption level (Parfitt et al. 2010), which is the result of customers' excessive consumption habits or high food selection criteria (Kantor et al. 1997; Lipinski et al. 2013). The FAO defines it as "food appropriate for human consumption being discarded, whether or not after it is kept beyond its expiry date or left to spoil" (FAO 2013). It occurs mainly in developed countries (Gustavsson et al. 2011).

This book falls under the category of food loss, which occurs at the front end of the supply chain. 30%–40% of total food production is lost before it reaches the market (FAO 2022). Food loss is mostly driven by technical constraints such as inadequate harvesting techniques, poor infrastructure and logistics, a lack of storage facilities, and insufficient skills and management (FAO 2013; Lipinski et al. 2013; Neff et al. 2015). Therefore, less developed countries suffer more food loss than food waste (Gustavsson et al. 2011). Currently, there is a lack of consensus on the definition of food loss (Minor et al. 2020). Table 1.2 lays out some food loss definitions. In the food loss definitions of the United States Department of Agriculture (USDA) and the World Resources Institute (WRI), food waste is considered a subset of food loss. FAO, the Committee on World Food Security (CFS), and the United States Environmental Protection Agency (U. S. EPA) make a distinction between food loss and food waste. Although these three definitions remain different, they all exclude the consumer stage.

The definition by U.S. EPA (2020) clearly stated that food loss includes unharvested crops. Minor et al. (2020) also indicated that food that has matured but is left in the field unharvested is also a food loss. Some studies habitually use "post-harvest", but they did not exclude losses that occur during harvest. It is clear that harvest loss is a type of food loss that occurs at the harvest stage.

Before defining harvest loss, it is necessary to clarify the harvest stage. The operations performed during the harvest stage are not single. Harvest operations mainly occur in the field and consist of a series of operations such as reaping the panicles or cutting the stalks and bundling for transportation (Bala et al. 2010). The use of combine harvesters weakens the delineation between the post-processing stages of rice (i.e., reaping, threshing, and winnowing) compared to manual operations, making it difficult to strictly distinguish these operations (Wu et al. 2017; Qu et al. 2021).

Table 1.2 Definitions of food loss and food waste

Agency	Food loss	Food waste	Is waste included in loss?
FAO (2013)	"…a decrease in mass (dry matter) or nutritional value (quality) of food that was originally intended for human consumption"	"…food appropriate for human consumption being discarded, whether or not after it is kept beyond its expiry date or left to spoil"	No
CFS (Timmermans et al. 2014)	"…a decrease, at all stages of the food chain prior to the consumer level, in mass, of food that was originally intended for human consumption, regardless of the cause"	"…food appropriate for human consumption being discarded or left to spoil at consumer level — regardless of the cause"	No
USDA (Buzby et al. 2014)	"…the amount of edible food, post-harvest, that is available for human consumption but is not consumed for any reason"	NA	Yes
WRI (Hanson et al. 2016)	"…the weight of food and/or associated inedible parts removed from the food supply chain"	NA	Yes
U. S. EPA (2020)	"…unused product from the agricultural sector, such as unharvested crops"	"…food such as plate waste (i.e., food that has been served but not eaten), spoiled food, or peels and rinds considered inedible"	No

Note NA = not applicable
Sources See column 1 in table

Therefore, in this book, the harvest stage is defined as the process from reaping to on-farm storage places, which specifically includes four operations: reaping, threshing, winnowing, and field transport. Accordingly, harvest losses are the losses that occur during these four operations. Reaping or cutting is done by knife, serrated sickle, paddy mower, reaper, or combine harvester (Anujuprana et al. 2013). When reaping the panicles or cutting the straws, grains may be scattered across the field, plowed into the soil, and eaten by birds and rodents (Parfitt et al. 2010), which form the reaping loss. Threshing is to peel the paddy kernels or grains from the panicles (Bala et al. 2010; Sawicka 2019). This process is to apply external impact force on panicles, which can be done manually by trampling, banging with stick or flail device, pedal thresher, or by machinery such as power thresher and combine harvester (Bala et al. 2010; Anujuprana et al. 2013). During this process, grains remained on the panicles or scattered on the threshing floor constitute the threshing loss (Qu et al. 2021). After threshing, the immature grains, rice straw, stones, sand, chaff, and other foreign materials are removed from the threshed paddy by sieving, wind, or hand picking (Baloch 1999), during which paddy may also be removed, resulting in winnowing loss (Danbaba et al. 2019). Field transport refers to transferring the grain from the field or winnowing places to storage places. Field transport loss during this process may be caused by spillage (Sawicka 2019; Qu et al. 2021).

Moreover, the definition given by FAO (2013) indicates that the losses during harvesting include quantity and quality loss. Since the measurement of quality loss is more difficult than that of quantity loss (Taiwo and Bart-Plange 2016), and the lack of sufficient quality awareness to classify rice before it reaches the formal market (Hodges et al. 2014), this book focuses only on quantity loss. In summary, harvest losses in this book refer to quantity losses that occur between field reaping and on-farm storage places, specifically reaping loss, threshing loss, winnowing loss, and field transport loss.

1.3.2 Harvest Outsourcing Service

The concept of outsourcing service first appeared in a study named *Enterprise's core competitiveness*. Subsequently, outsourcing service has emerged in the field of agricultural production, which is also well-known as "farm contractor system". It is worth noting that the agricultural outsourcing service discussed here is different from the agricultural outsourcing between countries, which occurs when capital-rich countries buy or lease land in capital-poor countries for their domestic agricultural production (Hofman and Ho 2012; Vandergeten et al. 2016). Agricultural outsourcing service is the practice of outsourcing some or all stages of the agricultural production process to organizations or individuals that specialize in providing production services (Ji et al. 2017; Cai and Wang 2021).

Agricultural machinery outsourcing service is the earliest form and important content of agricultural outsourcing service. In China, agricultural machinery outsourcing service is also known as "agricultural machinery socialization service", "agricultural machinery leasing service", and "contract hire system of agricultural machines", among which, "agricultural machinery socialization service" is often used in official documents. In this book, the concept of agricultural machinery outsourcing service is adopted for the convenience of understanding, which can reveal its essence better. According to the Ministry of Agriculture and Rural Affairs of China (MARA), agricultural machinery outsourcing services refer to various agricultural machinery operation services such as mechanical plowing, mechanical sowing, mechanical harvesting, mechanical irrigation and drainage, as well as other related paid services such as agricultural machinery maintenance, supply, intermediation, and leasing provided by agricultural machinery service organizations or individuals to other agricultural producers (MARA, 2013).

Therefore, based on the definition by MARA, harvest outsourcing service studied in this book is a type of service that belongs to the agricultural machinery outsourcing service, as shown in Fig. 1.1. Specifically, harvest outsourcing service is that farmers outsource harvesting operations to agricultural machinery organizations or individuals and pay them service fees. The point that needs to be noted is that, in this process,

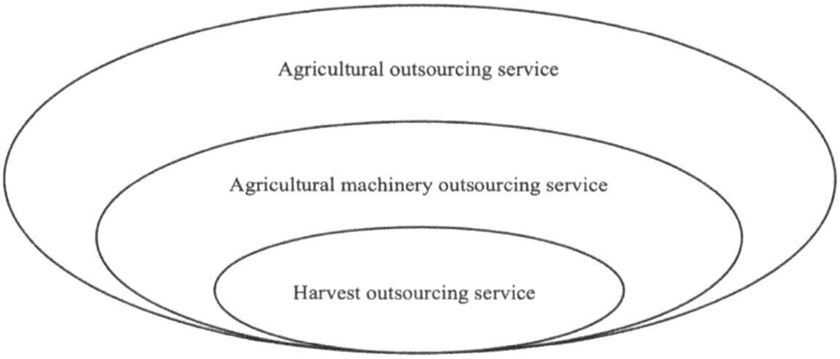

Fig. 1.1 Harvest outsourcing service

the machinery operators come from service organizations or individuals other than household members (Deng et al. 2020).

1.3.3 Moral Hazard

The term "moral hazard" originated in the fire insurance literature and appeared in the mid-nineteenth century (Ducat 1865). Moral hazards argue that purchasing insurance will induce individuals to reduce preventive inputs (Rowell and Connelly 2012). It has been described as "Men who would ordinarily scorn to steal or lie, will magnify a slight injury, or be dilatory in resuming work when they are able to do so" (McNeill 1900).

In the economics literature, moral hazard is widely used in principal–agent relationships. Adam Smith wrote in the *Wealth of Nations* that "The directors of such companies, however, being the managers rather of other people's money than of their own, it cannot well be expected that they should watch over it with the same anxious vigilance with which the partners in a private copartnery frequently watch over their own…Negligence and profusion, therefore, must always prevail, more or less, in the management of the affairs of such a company" (Smith 1937). Moral hazards occur when principals are unable to observe, or find the cost of observation inputs too high (Shavell 1979). Arrow (1984) defined moral hazard simply as the agents' hidden actions, where the effort of agents is the most typical one. Vaughan (1997) stated the moral hazard "…results from the insured person's careless attitude toward the occurrence of losses. The purchase of insurance may create a moral hazard since the realization that the insurance company will bear the loss".

In a nutshell, moral hazard is essentially the idea that the information advantaged party does not have to bear the costs of their actions (Silva and Yoshitomi 2001), thereby reducing their incentive to guard against risks, such as dishonesty, carelessness, and lack of attention (Rowell and Connelly 2012), which can lead to an increased probability of loss (Surminski 2013). Although we know that the information advantaged party will reduce their level of effort, it is difficult to determine whether they are committing moral hazards by the intensity of effort because the original intensity of effort cannot be observed. However, as Adam Smith said, the agents will not be as attentive as managing his own money (Smith 1937). Therefore, moral hazard can be simply described as a situation in which the information advantaged party is not as diligent or conscientious in reducing the occurrence of loss as the information disadvantaged party. Specifically for this book, the moral hazards of service providers are reflected in the fact that they reduce their own effort level in providing harvesting services, which is lower than that of farmers.

1.4 Structure of the Book

To meet the above objectives, this book is organized as follows (Fig. 1.2). This chapter provides the background of this book. This chapter describes the need and urgency of reducing rice harvest losses. It also describes the prevalence of harvest outsourcing services in China and their possible negative effect on rice harvest losses due to moral hazards. Chapter 2 reviews the existing research on rice harvest losses and harvest outsourcing services, with a focus on rice harvest losses. This chapter highlights the shortcomings of existing studies and the contributions of this study to literature. Chapter 3 provides a detailed description of the data used in this book. After introducing the two nationwide data used in this book, this chapter presents a detailed description of the estimation methods for the key variables and the statistical description of rice harvest loss. Chapter 4 begins by providing the evidence of moral hazards in harvest outsourcing services. Using Logit models and Propensity Score Matching method (PSM), the existence of moral hazards is studied through the comparison of work attitudes of service providers and farmers. Chapter 5 then studies whether the effect of harvest outsourcing services on operators' work attitudes will lead to increased rice harvest losses. This chapter examines the mediation effect of moral hazard using the mediation analysis model. In Chapter 6, Two-Stage Least Squares (2SLS) are used to investigate the effect of moral hazards on rice harvest losses of farmers who used harvest outsourcing services. Finally, Chapter 7 concludes with the policy implications and future works.

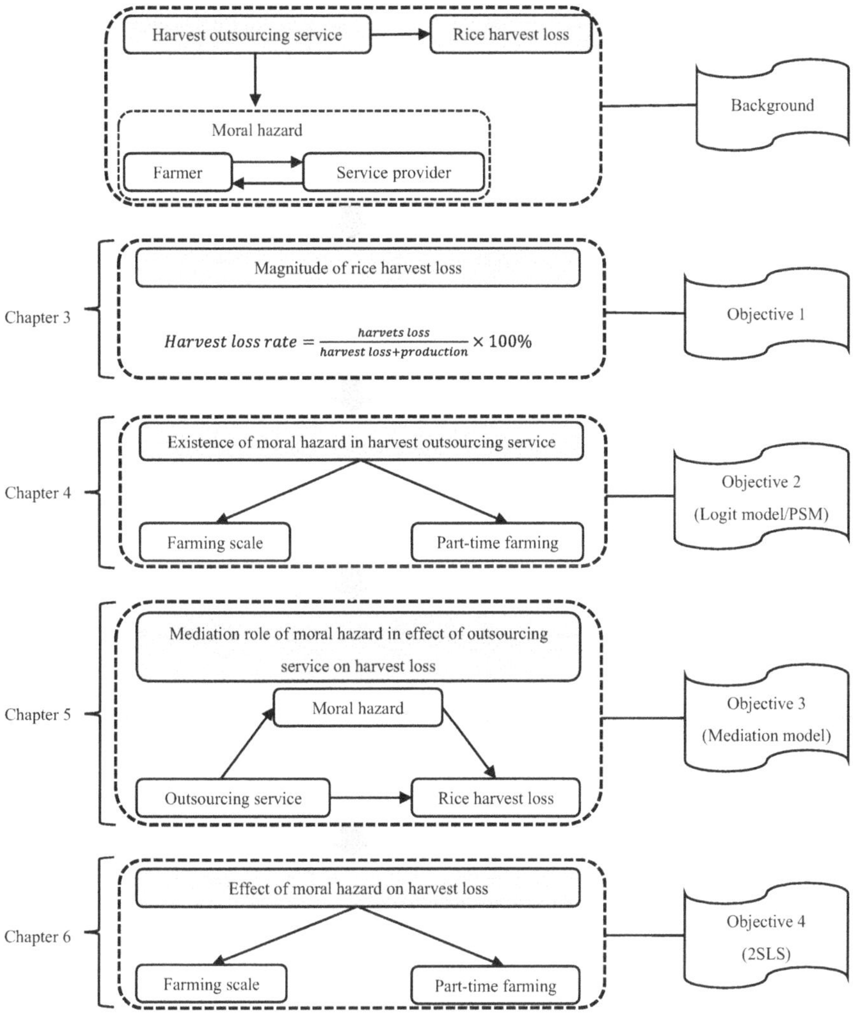

Fig. 1.2 Structure of the book

References

Affognon H, Mutungi C, Sanginga P, Borgemeister C (2015) Unpacking postharvest losses in Sub-Saharan Africa: a meta-analysis. World Dev 66:49–68. https://doi.org/10.1016/j.worlddev.2014.08.002

Andersson F, Jordahl H, Josephson J (2019) Outsourcing public services: contractibility, cost, and quality. Cesifo Econ Stud 65:349–372. https://doi.org/10.1093/cesifo/ifz009

Anujuprana AH, Machfud S, Suryani A (2013) Model for measuring post-harvest technological capability of paddy farmers in dealing with climate change. Innov Syst Des Eng 4:33–43. https://doi.org/10.1504/IJTPM.2014.060158

Arrow KJ (1984) The economics of agency. Stanford University institute for Mathematical Studies in the Social Sciences, Stanford

Arvanitoyannis IS, Tserkezou P (2008) Cereal waste management: treatment methods and potential uses of treated waste. Waste management for the food industries. Academic Press, Amsterdam, The Netherlands, pp 629–702

Babatunde R, Omoniwa A, Aliyu J (2019) Post-harvest losses along the rice value chain in Kwara State, Nigeria: an assessment of magnitude and determinants. Cercet Agron În Mold 52:141–150. https://doi.org/10.2478/cerce-2019-0014

Bai ZG, Dent DL (2006) Global assessment of land degradation and improvement: pilot study in Kenya. Wageningen, ISRIC-World Soil Information

Bala BK, Haque M, Hossain MA, Majumdar S (2010) Post harvest loss and technical efficiency of rice, wheat and maize production system: assessment and measures for strengthening food security. Bangladesh Agricultural University, Bengaluru, India

Baloch UK (1999) Wheat: post-harvest operations. Pakistan Agricultural Research Council, Islamabad

Barrera EL, Hertel T (2021) Global food waste across the income spectrum: implications for food prices, production and resource use. Food Policy 98:101874. https://doi.org/10.1016/j.foodpol.2020.101874

Basappa G, Deshmanya JB, Patil BL (2007) Post-harvest losses of maize crop in Karnataka—an economic analysis. Karnataka J Agric Sci 20:69–71

Belton B, Fang P, Reardon T (2018) Mechanization outsourcing services in Myanmar's dry zone. Michigan State University, Michigan, USA

Buzby JC, Wells HF, Hyman J (2014) The estimated amount, value, and calories of postharvest food losses at the retail and consumer levels in the United States. United States Department of Agriculture, Washington, DC

Cai R, Cai KS (2014) Empirical research on agricultural production outsourcing—based on the investigation of main rice producing areas in Anhui Province. J Agrotechnical Econ, 34–42

Cai J, Liu WY (2019) Agricultural social service and opportunistic behavior: take agricultural machinery operation services as example. Reform, 18–29

Cai LM, Wang LP (2021) Analysis on outsourcing service behavior of rice pest and disease control based on Heckman selection model—a case study of ten counties in Fujian Province. PLoS ONE 16:1–17. https://doi.org/10.1371/journal.pone.0254819

Chegere MJ (2017) Post-harvest losses, intimate partner violence and food security in Tanzania. University of Gothenburg, Gothenburg, Sweden

Chen XP, Cui ZL, Fan MS et al (2014) Producing more grain with lower environmental costs. Nature 514:486–489. https://doi.org/10.1038/nature13609

Coelli T, Battese G (1996) Identification of factors which influence the technical inefficiency of Indian farmers. Aust J Agric Econ 40:103–128. https://doi.org/10.1111/j.1467-8489.1996.tb00558.x

Coker AA, Ninalowo SO (2016) Effect of post-harvest losses on rice farmers' income in Sub-saharan Africa: a case of Niger state, Nigeria. J Agric Sci Food Technol 2:27–34

Danbaba N, Idakwo PY, Kassum AL et al (2019) Rice postharvest technology in Nigeria: an overview of vurrent status, constraints and potentials for sustainable development. Open Access Libr J 6:1–23. https://doi.org/10.4236/oalib.1105509

Delgado L, Schuster M, Torero M (2021) Quantity and quality food losses across the value Chain: a comparative analysis. Food Policy 98:101958. https://doi.org/10.1016/j.foodpol.2020.101958

Deng X, De XuD, Zeng M, Bin QY (2020) Does outsourcing affect agricultural productivity of farmer households? evidence from China. China Agric Econ Rev 12:673–688. https://doi.org/10.1108/CAER-12-2018-0236

Ducat AC (1865) The practice of fire underwriting, 4th edn. T. Jones

FAO (1981) Food loss prevention in perishable crops. Food and Agriculture Organization of the United Nations: Rome, Italy

FAO (2009) How to feed the world in 2050. https://www.fao.org/fileadmin/templates/wsfs/docs/expert_paper/How_to_Feed_the_World_in_2050.pdf#:~:text=. According to the latest revision of the UN,6.8 billion today to 9.1 billion in 2050. Accessed 23 Jan 2024

FAO (2013) Food wastage footprint: Impacts on natural resources. Food and Agriculture Organization of the United Nations, Rome, Italy

FAO (2015) United Nations sustainable development goals. Goal 12(3). https://www.fao.org/sustainable-development-goals/indicators/1231/en/. Accessed 17 Oct 2021

FAO (2022) Seeking end to loss and waste of food along production chain. https://www.fao.org/in-action/seeking-end-to-loss-and-waste-of-food-along-production-chain/en/. Accessed 7 Oct 2022

Foley JA, Ramankutty N, Brauman KA et al (2011) Solutions for a cultivated planet. Nature 478:337–342. https://doi.org/10.1038/nature10452

Gao LW, Xu SW, Li ZM et al (2016) Main grain crops postharvest losses and its reducing potential in China. Trans Chinese Soc Agric Eng 32:1–11

Godfray HCJ, Beddington JR, Crute IR, et al (2010) Food security: the challenge of feeding 9 billion people. Science 80(327):812–818. https://doi.org/10.1126/science.1185383

Greeley M (1982) Farm-level post-harvest food losses: the myth of the soft third option. IDS Bull 13:51–60. https://doi.org/10.1111/j.1759-5436.1982.mp13003007.x

Greeley M (1991) Postharvest technologies: implications for food policy analysis. Economic Development Institute of The World Bank, Washington, DC

Gustavsson J, Cederberg C, Sonesson U, et al (2011) Global food losses and food waste—Extent, causes and prevention. Food and Agricultural Organization of the United Nations: Rome, Italy

Han QL (2019) On the cohesion dilemma between family management and agricultural social service: based on M county. J Nanjing Agric Univ Social Sci Ed 19:20–27

Hanson C, Lipinski B, Robertson K et al (2016) Food loss and waste accounting and reporting standard. FLW Protocol Steering Committee

Hodges RJ, Buzby JC, Bennett B (2010) Postharvest losses and waste in developed and less developed countries: Opportunities to improve resource use. J Agric Sci 149:37–45. https://doi.org/10.1017/S0021859610000936

Hodges RJ, Bernard M, Rembold F (2014) APHLIS—postharvest cereal losses in Sub-Saharan Africa, their estimation, assessment and reduction. European Union, Luxembourg

Hofman I, Ho P (2012) China's "developmental outsourcing": a critical examination of Chinese global "land grabs" discourse. J Peasant Stud 39:1–48. https://doi.org/10.1080/03066150.2011.653109

Huan ML, Hou YX (2020) Quality control contract model of service in agricultural production outsourcing. J Agro-Forestry Econ Manag 19:288–296. https://doi.org/10.16195/j.cnki.cn36-1328/f.2020.03.31

Huang D, Yao L, Wu LP, Zhu X Di (2018) Measuring rice loss during harvest in China: based on experiment and survey in five provinces. J Nat Resour 33:1427–1438. https://doi.org/10.31497/zrzyxb.20170810

Ibrahim H, Saba S, Ojoko EA (2018) Post-harvest loss in rice production: evidence from a rural community in Northern Nigeria. FUDMA J Sci 2:17–22

Jalava M, Guillaume JHA, Kummu M et al (2016) Diet change and food loss reduction: What is their combined impact on global water use and scarcity? Earth's Futur 4:62–78. https://doi.org/10.1002/2015EF000327

Ji C, Guo HD, Jin SQ, Yang J (2017) Outsourcing agricultural production: evidence from rice farmers in Zhejiang province. PLoS ONE 12:1–16. https://doi.org/10.1371/journal.pone.0170861

Kantor LS, Lipton K, Manchester A, Oliveira V (1997) Estimating and addressing America's food losses. Food Rev Food Rev 20:2–12. https://doi.org/10.22004/ag.econ.234453

Kummu M, de Moel H, Porkka M et al (2012) Lost food, wasted resources: global food supply chain losses and their impacts on freshwater, cropland, and fertiliser use. Sci Total Environ 438:477–489. https://doi.org/10.1016/j.scitotenv.2012.08.092

Li M, Sicular T (2013) Aging of the labor force and technical efficiency in crop production: evidence from Liaoning province, China. China Agric Econ Rev 5:342–359. https://doi.org/10.1108/CAER-01-2012-0001
Li ZF, Xia PK, Wang ZH et al (1991) Analysis of the constitution of grain postproduction losses and the preventive measures. J Zhejiang Univ 17:389–395
Li B, Qian Y, Kong F (2023) Does outsourcing service reduce the excessive use of chemical fertilizers in rural China? the moderating effects of farm size and plot size. Agriculture 13:1869. https://doi.org/10.3390/agriculture13101869
Lipinski B, Hanson C, Lomax J, et al (2013) Reducing food loss and waste. World Resources Institute, Washington, DC
Liu JG, Lundqvist J, Weinberg J, Gustafsson J (2013) Food losses and waste in China and their implication for water and land. Environ Sci Technol 47:10137–10144. https://doi.org/10.1021/es401426b
Lu QA, Du XD (2020) The outsourcing choice of agricultural production tasks:Implications for food security—a multiple-task based approach. In: The 2020 Agricultural & Applied Economics Association Annual Meeting. Kansas City, Missouri
Lu S, Cheng G, Li T et al (2022) Quantifying supply chain food loss in China with primary data: a large-scale, field-survey based analysis for staple food, vegetables, and fruits. Resour Conserv Recycl 177:106006. https://doi.org/10.1016/j.resconrec.2021.106006
Maclean J, Hardy B, Hettel G (2013) Rice Almanac: source book for one of the most important economic activity on earth, 4th edn. International Rice Research Institute, Los Baños, Philippines
MARA (2013) Ministry of Agriculture on vigorously promoting the socialization of agricultural machinery services. http://www.moa.gov.cn/nybgb/2013/dshiq/201712/t20171219_6121447.htm. Accessed 15 Oct 2021
MARA (2018) The Ministry of Agriculture and Rural affairs promoted the upgrading of socialized services for agricultural machinery. https://china.gov.cn.admin.kyber.vip/xinwen/2018-11/27/content_5343755.htm. Accessed 27 Dec 2021
McNeill GE (1900) A study of accidents and accident insurance. Insurance Topics Company, Boston
Mi Q, Li XD, Gao JZ (2020) How to improve the welfare of smallholders through agricultural production outsourcing: evidence from cotton farmers in Xinjiang. Northwest China. J Clean Prod 256:120636. https://doi.org/10.1016/j.jclepro.2020.120636
Minor T, Astill G, Skorbiansky SR, et al (2020) Economic drivers of food loss at the farm and pre-retail sectors: a look at the produce supply chain in the United States. United States Department of Agriculture
Müller A, Nunes MT, Maldaner V et al (2022) Rice drying, storage and processing: effects of post-harvest operations on grain quality. Rice Sci 29:16–30. https://doi.org/10.1016/j.rsci.2021.12.002
Neff RA, Kanter R, Vandevijvere S (2015) Reducing food loss and waste while improving the public's health. Health Aff 34:1821–1829. https://doi.org/10.1377/hlthaff.2015.0647
Parfitt J, Barthel M, MacNaughton S (2010) Food waste within food supply chains: quantification and potential for change to 2050. Philos Trans R Soc B Biol Sci 365:3065–3081. https://doi.org/10.1098/rstb.2010.0126
Picazo-Tadeo AJ, Reig-Martínez E (2006) Outsourcing and efficiency: the case of Spanish citrus farming. Agric Econ 35:213–222. https://doi.org/10.1111/j.1574-0862.2006.00154.x
Pierrat É, Laurent A, Dorber M, et al (2023) Advancing water footprint assessments: combining the impacts of water pollution and scarcity. Sci Total Environ 870. https://doi.org/10.1016/j.scitotenv.2023.161910
Pingali P (2007) Agricultural mechanization: adoption patterns and economic impact. Handb Agric Econ 3:2779–2805. https://doi.org/10.1016/S1574-0072(06)03049-0
Qian H, Zhu X, Huang S et al (2023) Greenhouse gas emissions and mitigation in rice agriculture. Nat Rev Earth Environ 4:716–732. https://doi.org/10.1038/s43017-023-00482-1

Qu X, Kojima D, Nishihara Y et al (2021) Can harvest outsourcing services reduce field harvest losses of rice in China? J Integr Agric 20:1396–1406. https://doi.org/10.1016/s2095-3119(20)63263-4

Ranganathan J, Waite R, Searchinger T, Hanson C (2018) How to sustainably feed 10 billion people by 2050. World Resour Inst. https://www.wri.org/insights/how-sustainably-feed-10-billion-people-2050-21-charts. Accessed 15 Jan 2022

Ren C, Zhou X, Wang C et al (2023) Ageing threatens sustainability of smallholder farming in China. Nature 616:96–103. https://doi.org/10.1038/s41586-023-05738-w

Rowell D, Connelly LB (2012) A history of the term "moral hazard." J Risk Insur 79:1051–1075. https://doi.org/10.1111/j.1539-6975.2011.01448.x

Rozelle S, Taylor JE, DeBrauw A (1999) Migration, remittances, and agricultural productivity in China. Am Econ Rev 89:287–291. https://doi.org/10.1257/aer.89.2.287

Ruttan VW (2000) Technology, growth and development: an induced innovation perspective. Oxford University Press, New York

Sawicka B (2019) Post-harvest losses of agricultural produce. Sustain Dev 1:1–16. https://doi.org/10.1007/978-3-319-69626-3_40-1

Shafiee-Jood M, Cai X (2016) Reducing food loss and waste to enhance food security and environmental sustainability. Environ Sci Technol 50:8432–8443. https://doi.org/10.1021/acs.est.6b01993

Shavell S (1979) On moral hazard and insurance. Q J Econ, 541–562. https://doi.org/10.1007/978-94-015-7957-5_15

Shen HF, Chen C, Liao XY, Wang L (2015) Spatial econometric analysis of rice production links outsourcing pricing mechanism: empirical study of 14 province 42 counties in China. China Rural Surv, 34–46

Sheng Y, Song LG, Yi Q (2017) Mechanisation outsourcing and agricultural productivity for small farms: implications for rural land reform in China. In: Song L, Garnaut R, Cai F, Johnston L (eds) China's new sources of economic growth: human capital, innovation and technological change, vol 2. Australian National University. Acton, Australia, pp 289–313

Silva LAP da, Yoshitomi M (2001) Can "moral hazard" explain the Asian crises? Asian Development Bank Institute, Tokyo, Japan

Smith A (1937) The wealth of nations: an inquiry into the nature and causes of the wealth of nations. University of Oregon, Eugene, USA

Sun SK, Lu YJ, Gao H et al (2018a) Impacts of food wastage on water resources and environment in China. J Clean Prod 185:732–739. https://doi.org/10.1016/j.jclepro.2018.03.029

Sun DQ, Rickaille M, Xu ZG (2018b) Determinants and impacts of outsourcing pest and disease management: evidence from China's rice production. China Agric Econ Rev 10:443–461. https://doi.org/10.1108/CAER-01-2017-0011

Sun XH (2013) Agricultural production operator: Comparison of types and path choice—from the ferspective of the overall production efficiency. Res Econ Manag, 59–66. https://doi.org/10.13502/j.cnki.issn1000-7636.2013.12.007

Surminski S (2013) The role of insurance in reducing direct risk—the case of flood insurance. Int Rev Environ Resour Econ 7:241–278. https://doi.org/10.1561/101.00000062

Taiwo A, Bart-Plange A (2016) Factors responsible for post-harvest losses and their effects on rice producing farmers: a case study of Afife and Aveyime rice projects in the Volta region of Ghana. Int Res J Eng Technol 3:1014–1022

Tan ZP, Hong WJ, Luo BL (2019) The transfer effect of agricultural labor force and grain-oriented planting structure. Reform, 111–118

Tefera T (2012) Post-harvest losses in African maize in the face of increasing food shortage. Food Secur 4:267–277. https://doi.org/10.1007/s12571-012-0182-3

Timmermans AJM, Ambuko J, Belik W, Huang J (2014) Food losses and waste in the context of sustainable food systems. A report by the High Level Panel of Experts on Food Security and Nutrition of the Committee on World Food Security: Rome, Italy

Tocco B, Davidova S, Bailey A (2012) Key issues in agricultural labour markets: A review of major studies and project reports on agriculture and rural labour markets. Factor Markets research Project

U.S. EPA (2020) Advancing sustainable materials management: 2018 wasted food report. United States Environmental Protection Agency

Vandergeten E, Azadi H, Teklemariam D et al (2016) Agricultural outsourcing or land grabbing: a meta-analysis. Landsc Ecol 31:1395–1417. https://doi.org/10.1007/s10980-016-0365-y

Vaughan EJ (1997) Risk management. John Willey, New York

Wang ZG, Shen HF, Liao XY (2011) Agricultural scale management: starting from outsourcing of production links—a case study of rice. Chinese Rural Econ, 4–12

Wang XB, Yamauchi F, Huang JK (2016) Rising wages, mechanization, and the substitution between capital and labor: evidence from small scale farm system in China. Agric Econ 47:309–317. https://doi.org/10.1111/agec.12231

Wang R, Zhang Y, Zou C (2022) How does agricultural specialization affect carbon emissions in China? J Clean Prod 370:133463

Wei S, Lu Y (2024) Adoption mode of agricultural machinery and food productivity: evidence from China. Front Sustain Food Syst 7. https://doi.org/10.3389/fsufs.2023.1257918

Wu LH, Hu QP, Wang JH, Zhu D (2017) Empirical analysis of the main factors influencing rice harvest losses based on sampling survey data of ten provinces in China. China Agric Econ Rev 9:287–302. https://doi.org/10.1108/CAER-03-2016-0036

Xie HL, Lu H (2017) Impact of land fragmentation and non-agricultural labor supply on circulation of agricultural land management rights. Land Use Policy 68:355–364. https://doi.org/10.1016/j.landusepol.2017.07.053

Xu D, Deng X, Guo SL, Liu SQ (2019) Labor migration and farmland abandonment in rural China: empirical results and policy implications. J Environ Manage 232:738–750. https://doi.org/10.1016/j.jenvman.2018.11.136

Xu X, Sharma P, Shu S et al (2021) Global greenhouse gas emissions from animal-based foods are twice those of plant-based foods. Nat Food 2:724–732

Yang FT (2007) Analysis on the principal–agent relationship between agricultural machinery outsourcer and agricultural machinery operation service provider. Chinese Agric Mech, 15–18. https://doi.org/10.13733/j.jcam.issn.2095-5553.2007.01.004

Yi Q (2018) Adoption of agricultural mechanization services among maize farmers in China: Impacts of population aging and off-farm employment. In: 30th International Conference of Agricultural Economists. Vancouver

Yu Y, Jaenicke EC (2020) Estimating food waste as household production inefficiency. Am J Agric Econ 102:525–547. https://doi.org/10.1002/ajae.12036

Yue S, Xue Y, Lyu J, Wang K (2023) The effect of information acquisition ability on farmers' agricultural productive service behavior: an empirical analysis of corn farmers in Northeast China. Agriculture 13:573. https://doi.org/10.3390/agriculture13030573

Zhan YR (1995) Sampling survey and analysis of national grain post-harvest losses. Chian Grain Econ 4:44–47

Zhang XB, Yang J, Thomas R (2017) Mechanization outsourcing clusters and division of labor in Chinese agriculture. China Econ Rev 43:184–195. https://doi.org/10.1016/j.chieco.2017.01.012

Chapter 2
Literature Review

In this chapter, rice harvest losses are reviewed in terms of estimation methods, magnitudes of loss, causes of loss, impacts, and interventions. Existing research on harvest outsourcing services is also reviewed. Based on the review of existing studies, the contributions of this book to the existing literature are summarized.

2.1 Rice Harvest Loss

The research on rice harvest losses is reviewed by accessing both peer-reviewed articles and national and international reports (e.g., FAO and World Bank) through multiple major databases (e.g., Web of Science, Google Scholar, Scopus, Semantic Scholar, and China National Knowledge Infrastructure), though peer-reviewed articles are preferred. Searched keywords included "food loss" OR "food waste" OR "grain loss" OR "grain waste" OR "crop loss" OR "crop waste" OR "rice loss" OR "harvest loss" OR "post-harvest loss" OR "lost food". Strictly speaking, losses during harvesting are post-production rather than post-harvest losses. However, harvesting is often considered the starting point of the post-harvest management process, and existing studies do not strictly distinguish these stages, so we use "post-harvest" for retrieval. In addition, we perform supplemental searches for additional sources by reviewing article bibliographies.

Articles are screened for eligibility using the following inclusion criteria: (1) published in English or Chinese; (2) published in the social sciences rather than the natural sciences or engineering; (3) estimated the food loss rather than the food waste; (4) estimated the rice harvest loss (or reaping loss, threshing loss, winnowing loss, and field transportation loss) or corresponding causes, impacts, and interventions. After removal of duplicates and the above four inclusion criteria, unrelated articles are excluded by scanning the titles, abstracts, and full-text reading.

X. Qu et al., *Harvest Loss in China*, The University of Tokyo Studies on Asia,
https://doi.org/10.1007/978-981-97-9156-9_2

2.1.1 Estimation Methods

The first step to reduce losses is to estimate the magnitudes in each stage. Post-harvest losses in rice may be quantitative or qualitative. Quantitative losses result in weight or volume reduction in the potential yield, while qualitative losses result in a reduction in the value of the usable rice owing to physical and chemical changes, such as poor appearance, poor taste, and unpleasant odor (Babatunde et al. 2019). Nonetheless, harvest losses are usually measured as actual physical losses (i.e., quantitative losses). This is partly because qualitative losses are more difficult to measure than quantitative ones (Taiwo and Bart-Plange 2016), and partly because there is often not enough quality awareness to distinguish between grades of rice before it reaches the formal market (Hodges et al. 2014).

There are two main approaches to estimate harvest losses in existing studies: direct measurement and farmers' self-reported estimation (Hodges et al. 2010). Direct measurement of losses is usually done by conducting rice harvest experiments in the field and then collecting and weighing the lost rice (Gummert et al. 2020). Estimates from direct measurement may be more accurate but are time- and resource-consuming (Kitinoja et al. 2018). Comparatively, self-reported estimation is widely used (Delgado et al. 2021). Farmers experiencing losses are surveyed using a carefully designed structured questionnaire to elicit their estimates (Hodges et al. 2010). The accuracy of farmers' self-report data has been questioned compared to direct measurement. However, there is no actual evidence that farmers' measurement errors are larger than the errors in other estimates. Thus, farmer self-reports obtained based on well-designed questionnaires are considered reliable estimates (Sheahan and Barrett 2017). However, Kannan (2014) believed that farmer estimates may be subjective and are best validated by experts. In addition, some estimation methods are rarely used, such as modeling (Kitinoja et al. 2018) and Food Loss Analysis (FLA) methodology (a comprehensive methodology integrating literature review, expert interviews, questionnaires, observation and measurement, and load tracking and sampling) (FAO 2018; Totobesola et al. 2022).

Whether direct measurement or self-reported estimation, rice losses at each stage need to be clearly determined. Earlier studies on rice harvest losses in China appeared in the early 1990s (Li et al. 1991; Zhan 1995). Some new studies on rice harvest losses have emerged in the recent years (Gao et al. 2016; Wang et al. 2016; Wu et al. 2017; Huang et al. 2018), but these studies have different definitions of harvest losses, or lack of definitions, which makes comparisons difficult. Therefore, it is necessary to estimate the magnitude of the rice harvest losses under the clear definition. The rice losses estimated in reaping, threshing, winnowing, and field transport is discussed below.

2.1.1.1 Rice Loss During Reaping

Reaping is done by manual or mechanical methods, and the form of loss varies depending on the harvest method. There are usually two harvest methods: combine harvesting (i.e., head-feed and whole-feed) and segmented harvesting (manual reaping and reaper plus manual threshing or thresher) (Qu et al. 2021a). Alizadeh and Allameh (2013) referred to them as direct harvesting and indirect harvesting, respectively. When using segmented harvesting, the losses in the reaping stage include scattering loss and uncut loss (Li et al. 1991; Hodges et al. 2014; Amusat et al. 2016). Scattering loss occurs when rice grains fall in the field as a result of touching or other external forces owing to the influence of various aspects such as harvest method, crop variety, and maturity when cutting the straw or panicle. Uncut loss occurs when the straw or panicle is not cut because of careless working, lodging, and other reasons. To determine the scattering loss and uncut loss, grains on the ground and the uncut straw or panicle in the sample frame (1 m × 1 m or other sizes) or experimented plots are collected and weighed (Badawi 2001; Alizadeh and Allameh 2013; Amusat et al. 2016). If the cut straw directly moves to the next threshing operation, then the reaping loss is limited to the description above. However, in some regions, the cut straw is laid in the field for a few days before threshing and bundled for transportation to the homestead or threshing yard (FAO 2018). The grains that fall into the field during these days also count as the loss during reaping, or, in other words, stacking loss and bundling loss (Grolleaud 2002; Jha et al. 2015). To assess these losses, plastic sheets are placed under the grain stacks, and the grain that falls on the sheets is collected later for counting (Badawi 2001; Hodges et al. 2014; Nath et al. 2016).

When using combine harvesters, there is no stacking loss or bundling loss—only scattering loss and uncut loss. Since combine harvesters complete the following threshing and winnowing operations at once, the loss here also includes the threshing and winnowing losses described below (Alizadeh and Allameh 2013). Although Hasan et al. (2019) considered combine harvesting loss to also include cutter bar loss and cylinder loss, they only estimated the total loss. After the completion of combine harvesting, suitable sample frames (1 m × 1 m or other sizes) are randomly selected in the harvested fields. To estimate the loss, the dropped grains in the field and uncut straws from the sample area are collected and weighed for counting (Alizadeh and Allameh 2013; Jha et al. 2015).

2.1.1.2 Rice Loss During Threshing

Threshing could also be done by manual or mechanical methods. Manual threshing is done by the grain flail, threshing board or rack (Bala et al. 2010), bag-beating, "bam-bam" (a wooden box) (Guisse 2010), and pedal thresher (Anujuprana et al. 2013). Mechanical threshing employs power threshers and combine harvesters (Anujuprana et al. 2013; Sanneh 2015). Regardless of the method, losses during threshing are broadly divided into scattering loss and unthreshed loss (Hodges et al. 2014; Amusat et al. 2016). Manual or mechanical efforts to separate paddy from panicles can result in scattered or spilled rice, leading to scattering loss. After threshing, the rice grains that fall outside the plastic sheets or wooden boxes are collected and weighed (Guisse 2010; Amusat et al. 2016). Unthreshed loss refers to the grain remaining on the seed head. Assessment of this remaining grain can be done by taking a random sample of straws after threshing and separating the grains on the straws and weighing them (Greeley 1982; Hodges et al. 2014; Amusat et al. 2016; Nath et al. 2016).

2.1.1.3 Rice Loss During Winnowing

Loss during winnowing refers to the grain discarded with the straw and external impurities. A suitable sample of straws after winnowing is taken, and the grains blown away with the straw are separated and weighed (Jha et al. 2015).

2.1.1.4 Rice Loss During Field Transport

Losses during field transport from the field to the on-farm storages are mainly owing to broken packaging, which causes grains to fall to the ground. Measuring losses during field transport requires careful collection of scattered grains or weighing of grain bags at both the start and end of the transport process (Badawi 2001; Hodges et al. 2014).

2.1.2 Magnitude of Rice Harvest Loss

It is difficult to present a definite figure on rice harvest losses owing to differences in regions, harvest methods, and varieties. Even in the same region, there can be significant differences by season. Therefore, it is not realistic to describe rice harvest losses with a single figure. The estimation of losses then requires a situation-specific analysis (FAO 2018). Thus, rather than attempting to derive an exact loss figure from the reviewed studies, we attempt to determine some characteristics of these data.

Table 2.1 presents the ranges of rice harvest losses for some countries and regions. Most countries and regions have a harvest loss rate of less than 10%. Quantitative loss of rice harvest measured in Egypt is lowest, 1.35–2.49%. In Democratic Republic of Congo, Myanmar (wet season), and Dominican Republic, rice harvest loss rates can exceed 20%. In China, rice harvest loss rate is 1.23–5.5%.

Among the studies reviewed, only one was conducted in North America (Boxall et al. 1981), while the rest examined the Asian and African regions, with more than 70% focused on Asia. Most studies expressed the magnitude of quantitative loss as a percentage, and a few used the absolute weight of loss (Candia et al. 2012; Bordoloi 2013; Sarkar et al. 2013; Coker and Ninalowo 2016) and market value (Sanneh 2015; FAO 2018; Danbaba et al. 2019), while only one study measured qualitative losses

Table 2.1 Rice harvest losses in some countries and regions

Countries or regions		Harvest losses	Citation(s)
Africa	Democratic Republic of Congo	Quantitative loss: 23%	(Totobesola et al. 2022)
	Egypt	Quantitative loss: 1.35–2.49%	(Badawi 2001)
	Ghana	Quantitative loss: 3.57–16.14% Economic loss: 64.79 GH¢	(Guisse 2010; Appiah et al. 2011; Sanneh 2015; Amponsah et al. 2018)
	Nigeria	Quantitative loss: 4.84–9.73% Economic loss: 230.11 billion naira	(Oguntade et al. 2014; Amusat et al. 2016; Coker and Ninalowo 2016; Danbaba et al. 2019)
	Sub-Sahara Africa	Quantitative loss: 7.9–13.1%	(Ndindeng et al. 2021)
	Uganda	Quantitative loss: 11.1–15.8% (224.7–337.8 kg/ha)	(Candia et al. 2012)
Asia	Bangladesh	Quantitative loss: 1.61–6.95%	(Greeley 1982; Bala et al. 2010; Begum et al. 2012; Nath et al. 2016; Alam et al. 2018; Hasan et al. 2019)
	China	Quantitative loss: 1.23–5.5%	(Li et al. 1991; Grolleaud 2002; Gao et al. 2016; Wu et al. 2017; Huang et al. 2018; Gu et al. 2020; Qu et al. 2021a, b)
	Democratic Republic of Timor-Leste	Quantitative loss: 10.15% Economic loss: USD 9100	(FAO 2018)
	India	Quantitative loss: 1.60–5.95 kg/quintal; 2.88–3.60%	(Basavaraja et al. 2007; Veerangouda et al. 2010; Grover et al. 2012; Bordoloi 2013; Kannan et al. 2013; Roy 2013; Sarkar et al. 2013; Sivagnanam 2013; Jha et al. 2015)
	Indonesia	Quantitative loss: 8.26–8.83%	(Gaiser and Esmay 1981)
	Iran	Quantitative loss: 2.26–2.58% Qualitative loss: 0.47–2.44%	(Alizadeh and Allameh 2013)
	Myanmar	Quantitative loss: 16.0–28.2% (wet season); 0.9–9.3% (dry season)	(Grolleaud 2002; Gummert et al. 2020)

(continued)

Table 2.1 (continued)

Countries or regions		Harvest losses	Citation(s)
	Thailand	Quantitative loss: 1.1–9.3%	(Grolleaud 2002)
North America	Dominican Republic	Quantitative loss: 12.27–24.82%	(Boxall et al. 1981)

Note Countries or regions are sorted alphabetically. See Table A.7 in Appendices for more details
Sources See column 4 in the table

(Alizadeh and Allameh 2013). The losses in reaping and threshing are larger than those in winnowing and field transport (Table A.7 in Appendices) (Oguntade et al. 2014; Amusat et al. 2016).

Preliminary data obtained from rice harvesting trials indicate absolute loss in terms of the actual weight of grains. A more common form of quantitative loss is percentage loss (Sadiya and Hassan 2018), which makes comparisons easier (Qu et al. 2021a). However, different studies calculated the loss percentage with different denominators. Some used the remaining amount in the previous stage as the denominator (Sadiya and Hassan 2018), while others chose the final weight of collected grains at the present stage as the denominator (Guisse 2010; Amusat et al. 2016). In some, potential yield—the sum of final collected grain and losses—was used as the denominator (Qu et al. 2021a). However, even if the expressions for losses are the same, losses from different studies, conducted in different regions, seasons, and varieties, are hard to be compared (Guisse 2010).

2.1.3 Causes of Rice Harvest Loss

Some studies used questionnaires to directly inquire farmers' understanding of the causes of rice harvest losses (Kong et al. 2015; Coker and Ninalowo 2016; Taiwo and Bart-Plange 2016; Amponsah et al. 2018). Multiple regression analyses were also applied to estimate the influencing factors of the losses (Basavaraja et al. 2007; Begum et al. 2012; Wu et al. 2017; Qu et al. 2021a, b). Factors that contribute to harvest losses include harvest practice, environmental, socio-economic, and mechanical (Amponsah et al. 2018).

Harvesting time is one of the most important aspect of crop productions: early or delayed harvesting can lead to increased losses (Grover et al. 2012; Roy 2013; Kebede et al. 2019). If harvested too early, immature grains can make threshing difficult, resulting in more unthreshed loss (Wang et al. 2016). Delayed harvesting may also cause substantial loss owing to shattering and extended exposure to natural incidents, such as attacks by birds and other pests (Guisse 2010; Hasan et al. 2019; Gummert et al. 2020), which is in line with farmers' perspectives (Taiwo and Bart-Plange 2016). Regression analyses in China confirmed that harvesting in time was important for loss reduction (Wu et al. 2017).

A large family could reduce the drudgery involved in harvesting activities and thus reduce harvest losses (Adeola 2020). Manual harvesting is highly labor-intensive and tedious (Ssebaggala et al. 2017). If there is not enough labor or access to proper harvesting machinery, the mature rice cannot be harvested in time, causing massive losses (Zorya et al. 2011; FAO 2018; Kebede et al. 2019; Sawicka 2019; Gummert et al. 2020). The significance of this factor was also confirmed in the regression analysis (Basavaraja et al. 2007; Begum et al. 2012; Wu et al. 2017; Qu et al. 2021a).

Farm size affects the magnitude of losses. Some studies concluded that the average harvest loss was smaller for large-scale farmers than for small-scale farmers in China (Wu et al. 2017; Qu et al. 2021b), Bangladesh (Begum et al. 2012), Karnataka (Kannan et al. 2013), and West Bengal (Sarkar et al. 2013). However, the opposite results have also been observed in India (Basavaraja et al. 2007).

It was evident from the available research data that losses vary with rice varieties. High-yield varieties incur greater losses than local varieties in reaping, threshing, and winnowing (Bordoloi 2013), causing an unfortunate trade-off for farmers (Sheahan and Barrett 2017). High yields increase the pressure of reaping and threshing operations owing to increased yields, especially when there is a lack of machinery. Additionally, the thin husk and shells of high-yield rice make it more vulnerable to damage during the harvesting process (Greeley 1982).

Traditional manual operations are inefficient (Greeley 1986), which could cause higher losses. The losses due to the use of combine harvesters is smaller than that caused by manual reaping and threshers in Egypt (Badawi 2001). Similar results were found in Bangladesh (Hasan et al. 2019), Myanmar (Gummert et al. 2020), Dominican Republic (Boxall et al. 1981), and Thailand (Grolleaud 2002). However, some findings were contrary. In Bangladesh, new technology has been found to be associated with great losses (Greeley 1982). Amusat et al. (2016) found that mechanical operation causes higher losses at all three stages of reaping, threshing, and winnowing in Nigeria. In particular, threshing loss due to mechanical threshing is more than twice that caused by manual threshing. This is consistent with the findings of the questionnaire survey by Basavaraja et al. (2007). Several studies in China, including field trials (Li et al. 1991; Huang et al. 2018), household survey (Zhan 1995), multiple regression analyses (Li et al. 2020; Qu et al. 2021a), and a three-year International Development Research Centre survey (Grolleaud 2002), indicated that losses due to combine harvesting are higher than the total losses from segmented harvesting. Technological factors, maintenance, and driver' operation could be reason for high losses in mechanical harvesting (Kantor et al. 1997; Parfitt et al. 2010; Lu et al. 2022). Operators of combine harvesters may speed up the process to increase the area harvested per unit of time, leading to a high number of uncut plants and thus increasing the loss (Wu et al. 2017; Alam et al. 2018). Since combine harvesters cannot work effectively on land with excessive surface undulations, poor plot topography can greatly affect the work of the machinery (Taiwo and Bart-Plange 2016). Rice lodging can also make mechanical harvesting difficult and lead to increased losses (Huang et al. 2018). However, a rice harvesting trial in Iran by Alizadeh and Allameh (2013) showed that there was no significant difference in quantitative loss between combine harvesting and segmented harvesting, while the qualitative loss owing to broken, husked, and cracked grains was significantly less in combine harvesting. If the losses due to manual and mechanical operations are compared in one particular stage, the reaping loss caused by reaper use is greater than that by manual reaping in Bangladesh (Alam et al. 2018), and the threshing loss caused by mechanical threshing is higher than that by manual threshing in Nigeria (Amusat et al. 2016), India (Basavaraja et al. 2007), Bangladesh (Begum et al. 2012), and China (Grolleaud 2002). However, the winnowing loss due to mechanical winnowers is less than that due to manual operations (Kannan et al. 2013).

Even in manual operation, the loss varies with different methods. Field experiments showed that the reaping loss due to panicle harvesting (1.38%) was less than that due to sickle harvesting (2.93%) (Guisse 2010; Appiah et al. 2011). Threshing loss was higher using "bambam" than bagging (Guisse 2010; Appiah et al. 2011; Sanneh 2015). Field transport loss also varies by transportation method (Bordoloi 2013; Sarkar et al. 2013; Kannan 2014; Gao et al. 2016).

Losses are sometimes a result of farmers' choices. For example, harvest losses are often greater using machinery than manual operations. However, owing to the lack of sufficient agricultural labor, farmers will not forgo the use of machinery or other faster harvesting methods to reduce losses owing to the growing labor cost, especially for large-scale farmers (Guisse 2010). Soybean farmers in Brazil increase the forward speed of harvesters to maximize profits by harvesting earlier in the first season to leave time for the second season, although they are clearly aware of the higher losses involved (Goldsmith et al. 2015).

2.1.4 Impacts of Rice Harvest Loss

The most straightforward impact of rice harvest losses is a reduction in the amount of edible rice, resulting in economic losses and threatening poor people's livelihoods, especially in less developed countries (Ssebaggala et al. 2017; FAO 2018; Isatou and Sugh 2020). Farmers believe rice harvest losses threaten their food security and thus lead to poverty (Taiwo and Bart-Plange 2016). In Bangladesh, Begum et al. (2012) explored the effect of rice harvest loss on farmers' food security using logistic regression analysis and found that harvest loss has negative and significant relationships with households' probability of food security. Multiple regression analyses showed that rice threshing losses had a significantly negative effect on farmers' income in Nigeria (Coker and Ninalowo 2016).

The effect of these losses on the resource environment has been much mentioned. However, specific measurements of the effect on environmental and resource are relatively rare. In Nigeria, the loss of rice from production to milling results in the waste of 2.1 million m^3 of water, 0.5 million hectares of land, and the emission of about 0.65 million tons of carbon dioxide equivalent per year (Oguntade et al. 2014). The loss of rice from reaping to milling in Myanmar leads to an increase in greenhouse gas emissions of about 30–50% per hectare of rice production (Gummert et al. 2020).

2.1.5 Intervention Measures

Some common intervention measures targeting the reduction in rice harvest losses include improving infrastructure, introducing advanced machinery and equipment, timely harvesting, providing training to farmers, raising awareness of farmers on losses, and strengthening pest and disease management (Selvi et al. 2002; Basavaraja et al. 2007; Roy 2013; Sivagnanam 2013; Kannan 2014; Totobesola et al. 2022). However, proposing intervention measures is only the first step: What is more important is whether the intervention measures can be effectively implemented (Zorya et al. 2011; Rosegrant et al. 2015), whether they have significant effects (Rosegrant et al. 2015; Sheahan and Barrett 2017; Isatou and Sugh 2020; Stathers et al. 2020), and whether cost–benefit analysis presents important results (Rosegrant et al. 2015; Taiwo and Bart-Plange 2016; Sheahan and Barrett 2017).

Loss intervention measures should be designed from the perspective of farmers' profits. Whether farmers adopt new technologies and invest in loss reduction is primarily motivated by maximizing household profits, not production (Greeley 1982, 1986; Sheahan and Barrett 2017). Although some studies suggested that manual operation results in lower losses, this does not necessarily mean that manual operation should be encouraged, as delays in manual operations due to labor shortages may lead to even greater losses (Hasan et al. 2020). As farm labor becomes scarce, harvesters and threshers, or combine harvesting, have to be adopted. Thus, there is a trade-off between the need for mechanization and its associated higher losses (Guisse 2010). By quickly harvesting the first-season crop, Brazilian soybean growers gain more time for the second-season production (Goldsmith et al. 2015). If farmers are persuaded to reduce harvesting speed to minimize losses, they may not accept it because this would decrease their profitability. Similarly, hybrids and high-yield varieties are more likely to lead to post-harvest losses (Sheahan and Barrett 2017), but this does not mean farmers will abandon these seeds in order to reduce the loss.

Thus, the extent to which intervention measures will reduce losses is of concern. Gao et al. (2016) assumed that by increasing the proportion of combine harvesting, bulk transportation, mechanical drying, and depot storage in China to 100%, the total losses in rice reaping, transportation, drying, and storage can be reduced from 6.9 to 2.6%. Huang et al. (2018) assumed that if rice harvest loss rate by combine harvesting in China reaches 3%, the national rice harvest loss can be reduced from 3.02% to 2.76%, saving 540,000 tons of rice, 78,000 ha of land, and 26,000 tons of chemical fertilizer; if all the rice farmland is high standard type, the rice harvest loss can be further reduced to 2.08%. However, in addition to reducing losses, intervention measures may also have other effects. Rosegrant et al. (2015) found that increasing infrastructure investment is beneficial for reducing losses, improving social welfare, and has a positive economic rate of return. However, the increase in food supply leading to a drop in agricultural prices can result in loss of benefits for producers. The promotion of mechanical operations will replace labor (Stathers et al. 2020), thus potentially threatening the livelihoods of farmers who previously relied on manual post-harvest operations. In Bangladesh, the increasing use of mechanical rice milling can lead to the loss of jobs for women who were previously engaged in manual rice milling (Greeley 1991). Ensuring that the majority of participants benefit from mitigation strategies is a challenge in reducing food losses (Shafiee-Jood and Cai 2016).

It is essential to conduct a cost-benefit analysis of intervention measures, as no solution should be more expensive than the food losses themselves (Guisse 2010; FAO 2018). However, such analyses are rarely performed. Gummert et al. (2020) argued that increasing the level of mechanization can raise net income by 30–50%. Hasan et al. (2019) also indicated that using combine harvesters could save 57.61% in costs compared to manual harvesting; however, combine harvesters need to harvest over 35 hectares of land annually to be profitable. In the Democratic Republic of Timor-Leste, the savings from reduced losses over 10 years due to the introduction of threshers (approximately $33,200) exceeded the associated intervention cost of $14,000 (FAO 2018).

2.2 Harvest Outsourcing Service

Existing studies have examined harvest outsourcing services in terms of the origin and development of agricultural machinery outsourcing services, farmers' participation in outsourcing services and its influencing factors, and the effect of outsourcing services on agricultural production and farmers' welfare.

The aging of farm labor force due to low birth and death rates, the feminization of farm labor force caused by the male-dominated rural-to-urban migration driven by the wage differences between farm and non-farm employment, coupled with the changes in the relative prices of farm labor and harvest outsourcing services, have all contributed to the flourishing development of harvest outsourcing services in China. Harvest outsourcing services enable harvesting operations to be performed on a larger scale, thereby achieving economies of scale without altering the physical structure of the farms (Picazo-Tadeo and Reig-Martínez 2006; Zhang et al. 2017). Outsourcing services make the transformation from land scale operation to service scale operation, which is of great significance to the innovation of agricultural operation method (Luo 2014).

Farm size, labor endowments, government subsidies, and part-time farming can affect farmers' willingness to adopt outsourcing services (Igata et al. 2008; Cai and Cai 2014; Ji et al. 2017; Yi 2018). Compared to small-scale farms, large-scale farms are more interested in purchasing outsourcing services (Ji et al. 2017). Households with more family members working off-farm are more likely to use machinery outsourcing services (Yi 2018). Farmers with below-average productivity tend to purchase outsourcing services, while those with above-average productivity prefer to perform farming operations themselves (Deng et al. 2020).

Outsourcing services also have a positive effect on agricultural production. Outsourcing services strengthen the agricultural division of labor and generate economies of scale (Wolf 2003). On the one hand, machinery outsourcing services can improve agricultural productivity in China (Deng et al. 2020; Lu and Du 2020). A similar positive relationship between outsourcing services and production efficiency has been observed in Spanish citrus farming (Picazo-Tadeo and Reig-Martínez 2006). On the other hand, machinery outsourcing services contribute to cost saving (Tang et al. 2018). Additionally, outsourcing services improve farmers' welfare by increasing household income, boosting consumer spending, and improving labor accessibility (Mi et al. 2020).

A few studies have noted the moral hazard in agriculture outsourcing services from the perspective of principal-agent theory and game theory. Outsourcing services are essentially hired labor, which is less productive than household labor (Coelli and Battese 1996). Lu and Du (2020) argued that the moral hazard issues prevalent in outsourcing services of the secondary and tertiary industries are also present in agricultural outsourcing services. There are issues such as inconsistent goals and information asymmetries between farmers and service providers. Service providers may exhibit moral hazard behaviors, thus reducing production efficiency (Yang 2007; Huan and Hou 2020). However, these studies remain at the theoretical analysis stage.

2.3 Summary

This chapter reviews the existing literature on rice harvest losses and harvest outsourcing services. The existing literature has investigated harvest loss in terms of loss magnitude, causes of loss, effect of loss, and intervention measures. However, there are some limitations. Existing literature mainly focuses on quantitative losses and lacks research on qualitative losses. In China, the quantitative harvest loss rate of rice ranged from 1.23 to 5.5%. There is a lack of strict definitions for the stages at which losses are estimated, making comparison difficult. In terms of factors influencing rice harvest losses, existing studies have studied various factors from natural, human, and mechanical sources. A number of studies have discussed the effect of mechanical harvesting on harvest losses. In China, mechanical operations are thought to be associated with higher harvest losses, although contrary views exist. A common feature of these studies in China is that they all ignored the outsourcing services behind mechanical harvesting. In China, most mechanical harvesting is undertaken by harvest outsourcing services. A few studies have argued from theoretical analysis that there are moral hazard issues in harvest outsourcing services that may negatively affect on agricultural production (Huan and Hou 2020; Lu and Du 2020), such as increasing harvest losses. However, empirical research on this topic is currently lacking. There are many intervention measures proposed in the literature, such as improving infrastructure and increasing mechanization. However, few studies have analyzed the effectiveness and cost–benefit of these intervention measures.

Most studies on harvest outsourcing services focus on development backgrounds, farmers' willingness to participate and its influencing factors, and the effect of outsourcing services on agriculture and farmers. Only a few studies have noted the moral hazard in harvest outsourcing services and its potential negative impacts on agriculture, but they remain on theoretical analysis and lack empirical studies.

This study makes three contributions to the literature by estimating rice harvest losses in China and exploring the impact of moral hazard in harvest outsourcing service on rice harvest losses. Firstly, it provides the latest data of rice harvest losses from a nationwide survey of households since 1995. Secondly, it examines for the first time the influence of harvest outsourcing services, which has been overlooked in existing literature, on rice harvest losses in China. This can also be seen as a new perspective for evaluating the effect of outsourcing services on agriculture. Thirdly, using the mediation analysis, this study presents the first empirical research to verify that harvest outsourcing services have a negative effect on rice harvest losses through moral hazard. This provides empirical evidence for the existence of moral hazard in harvest outsourcing services.

References

Adeola EH (2020) Post-harvest management practices among rice farmers in Imo State Nigeria. Eur J Biol Biotechnol 1:1–6. https://doi.org/10.24018/ejbio.2020.1.4.32

Alam MA, Hossen A, Islam AS, Alam M (2018) Performance evaluation of power-operated reapers for harvesting rice at farmers' field. J Bangladesh Agric Univ 16:144–150. https://doi.org/10.3329/jbau.v16i1.36495

Alizadeh MR, Allameh A (2013) Evaluating rice losses in various harvesting practices. Int Res J Apllied Basic Sci 4:894–901

Amponsah SK, Addo A, Dzisi K et al (2018) Assessment of rice farmers' knowledge and perception of harvest and postharvest losses in Ghana. Cogent Food Agric 4:1471782. https://doi.org/10.1080/23311932.2018.1471782

Amusat MA, Eneh CK, Obiakor SC (2016) Assessment of postharvest losses of rice at different stages of operation. Int J Life Sci 5:50–53

Anujuprana AH, Machfud S, Suryani A (2013) Model for measuring post-harvest technological capability of paddy farmers in dealing with climate change. Innov Syst Des Eng 4:33–43. https://doi.org/10.1504/IJTPM.2014.060158

Appiah F, Guisse R, Dartey PKA (2011) Post harvest losses of rice from harvesting to milling in Ghana. J Stored Prod Postharvest Res 2:64–71

Babatunde R, Omoniwa A, Aliyu J (2019) Post-harvest losses along the rice value chain in Kwara State, Nigeria: an assessment of magnitude and determinants. Cercet Agron În Mold 52:141–150. https://doi.org/10.2478/cerce-2019-0014

Badawi AT (2001) A proposal on the assessment of rice post-harvest losses. In: The New Development in Rice Agronomy and Its Effects on Yield and Quality in Mediterranean Areas. Montpellier, CIHEAM

Bala BK, Haque M, Hossain MA, Majumdar S (2010) Post harvest loss and technical efficiency of rice, wheat and maize production system: assessment and measures for strengthening food security. Bangladesh Agricultural University, Bengaluru, India

Basavaraja H, Mahajanashetti SB, Udagatti NC (2007) Economic analysis of post-harvest losses in food grains in India: a case study of Karnataka. Agric Econ Res Rev 20:117–126. https://doi.org/10.22004/ag.econ.47429

Begum EA, Hossain MI, Papanagiotou E (2012) Economic analysis of post-harvest losses in food grains for strengthening food security in northern regions of Bangladesh. Int J Appl Res Bus Adm Econ 01:56–65

Bordoloi J (2013) Assessment of pre and post harvest losses of paddy and wheat in Assam. Agro-Economic Research Centre for North-East India Assam Agricultural University, Jorhat, India
Boxall RA, La Gra J, Martinez E, Martinez J (1981) Post harvest losses of rice in the Dominican Republic. Trop Stored Prod Inf 42:5–10
Cai R, Cai KS (2014) Empirical research on agricultural production outsourcing—based on the investigation of main rice producing areas in Anhui Province. J Agrotechnical Econ, 34–42
Candia A, Okurut S, Komaketch A et al (2012) On-farm post-harvest physical grain losses of "Kaiso" rice variety in Eastern Uganda. Uganda J Agric Sci 13:61–70
Coelli T, Battese G (1996) Identification of factors which influence the technical inefficiency of Indian farmers. Aust J Agric Econ 40:103–128. https://doi.org/10.1111/j.1467-8489.1996.tb00558.x
Coker AA, Ninalowo SO (2016) Effect of post-harvest losses on rice farmers' income in Sub-saharan Africa: a case of Niger state, Nigeria. J Agric Sci Food Technol 2:27–34
Danbaba N, Idakwo PY, Kassum AL et al (2019) Rice postharvest technology in Nigeria: an overview of vurrent status, constraints and potentials for sustainable development. Open Access Libr J 6:1–23. https://doi.org/10.4236/oalib.1105509
Delgado L, Schuster M, Torero M (2021) Quantity and quality food losses across the value Chain: a comparative analysis. Food Policy 98:101958. https://doi.org/10.1016/j.foodpol.2020.101958
Deng X, Xu D De, Zeng M, Qi Y Bin (2020) Does outsourcing affect agricultural productivity of farmer households? evidence from China. China Agric Econ Rev 12:673–688. https://doi.org/10.1108/CAER-12-2018-0236
FAO (2018) Food loss analysis: causes and solutions—case study on the rice value chain in the Democratic Republic of Timor-Leste. Food and Agriculture Organization of the United Nations, Rome, Italy
Gaiser D, Esmay M (1981) Traditional rice harvest loss and labor values in Indonesia. Trans ASAE 24:1162–1166. https://doi.org/10.13031/2013.34413
Gao LW, Xu SW, Li ZM et al (2016) Main grain crops postharvest losses and its reducing potential in China. Trans Chinese Soc Agric Eng 32:1–11
Goldsmith PD, Martins AG, de Moura AD (2015) The economics of post-harvest loss: a case study of the new large soybean—maize producers in tropical Brazil. Food Secur 7:875–888. https://doi.org/10.1007/s12571-015-0483-4
Greeley M (1982) Farm-level post-harvest food losses: the myth of the soft third option. IDS Bull 13:51–60. https://doi.org/10.1111/j.1759-5436.1982.mp13003007.x
Greeley M (1986) Food, thchnology and employment: the farm-level post-harvest system in developing countries. J Agric Econ 37:333–347. https://doi.org/10.1111/j.1477-9552.1986.tb01602.x
Greeley M (1991) Postharvest technologies: Implications for food policy analysis. Economic Development Institute of The World Bank, Washington, D.C.
Grolleaud M (2002) Post-harvest losses: discovering the full story Overview of the phenomenon of losses during the post-harvest system. Food and Agriculture Organization of United Nation, Rome, Italy
Grover DK, Singh JM, Singh P (2012) Assessment of pre and post harvest losses in wheat and paddy crops in Punjab. Agro-Economic Research Centre Department of Economics and Sociology Punjab Agricultural University, Ludhiana, India
Gu YN, Sun HY, Bi HW et al (2020) Effect of mechanical harvest on rice loss after mature in Heilongjiang. Agric Outlook 16:114–118
Guisse R (2010) Post harvest losses of rice (oriza spp) from harvesting to milling: a case study in Besease and Nobewam in the Ejisu Juabeng district in the Ashanti region of Ghana. Kwame Nkrumah University, Kabwe, Zambia
Gummert M, Nguyen-Van-Hung CC et al (2020) Assessment of post-harvest losses and carbon footprint in intensive lowland rice production in Myanmar. Sci Rep 10:1–13. https://doi.org/10.1038/s41598-020-76639-5

Hasan MK, Ali MR, Saha CK et al (2019) Combine harvester: impact on paddy production in Bangladesh. J Bangladesh Agric Univ 17:583–591. https://doi.org/10.3329/jbau.v17i4.44629
Hasan MK, Tanaka TST, Alam MM, et al (2020) Impact of modern rice harvesting practices over traditional ones. Rev Agric Sci 8:89–108. https://doi.org/10.7831/ras.8.0_89
Hodges RJ, Buzby JC, Bennett B (2010) Postharvest losses and waste in developed and less developed countries: opportunities to improve resource use. J Agric Sci 149:37–45. https://doi.org/10.1017/S0021859610000936
Hodges RJ, Bernard M, Rembold F (2014) APHLIS—postharvest cereal losses in Sub-Saharan Africa, their estimation, assessment and reduction. European Union, Luxembourg
Huan ML, Hou YX (2020) Quality control contract model of service in agricultural production outsourcing. J Agro-Forestry Econ Manag 19:288–296. https://doi.org/10.16195/j.cnki.cn36-1328/f.2020.03.31
Huang D, Yao L, Wu LP, Zhu X Di (2018) Measuring rice loss during harvest in China: Based on experiment and survey in five provinces. J Nat Resour 33:1427–1438. https://doi.org/10.31497/zrzyxb.20170810
Igata M, Hendriksen A, Heijman W (2008) Agricultural outsourcing: A comparison between the Netherlands and Japan. Appl Stud Agribus Commer 2:29–33. https://doi.org/10.19041/apstract/2008/1-2/4
Isatou J, Sugh ET (2020) Effect of extension dissemination on the control of post-harvest loss of rice in West Coast Region of The Gambia. Niger Agric J 51:22–28
Jha SN, Vishwakarma RK, Ahmad T et al (2015) Assessment of quantitative harvest and post-harvest losses of major crops/commodities in India. Ministry of Food Processing Industries, Ludhiana, India
Ji C, Guo HD, Jin SQ, Yang J (2017) Outsourcing agricultural production: Evidence from rice farmers in zhejiang province. PLoS One 12:1–16. https://doi.org/10.1371/journal.pone.0170861
Kannan E (2014) Assessment of pre and post harvest losses of important crops in India. Agricultural Development and Rural Transformation Centre, Institute for Social and Economic Change, Bangalore, India
Kannan E, Kumar P, Vishnu K, Abraham H (2013) Assessment of pre and post harvest losses of rice and red gram in Karnataka. Agricultural Development and Rural Transformation Centre, Bangalore, India
Kantor LS, Lipton K, Manchester A, Oliveira V (1997) Estimating and addressing America's food losses. Food Rev Food Rev 20:2–12. https://doi.org/10.22004/ag.econ.234453
Kebede L, Getnet B, Lema Y et al (2019) Advances in rice research and development in Ethiopia: post-harvest processes and advances to introduce loss-reducing technologies for rice. In: International Conference on Rice research and development organized at the inauguration of the Fogera National Rice Research & Training Center. Fogera, Ethiopia, pp 179–191
Kitinoja L, Tokala VY, Brondy A (2018) A review of global postharvest loss assessments in plant-based food crops: recent findings and measurement gaps. J Postharvest Technol 06:1–15
Kong S, Nanseki T, Chomei Y (2015) Farmers' perception of loss in post-harvest of rice yield in Cambodia. J Fac Agric Kyushu Univ 60:569–576. https://doi.org/10.5109/1543429
Li ZF, Xia PK, Wang ZH et al (1991) Analysis of the constitution of grain postproduction losses and the preventive measures. J Zhejiang Univ 17:389–395
Li XF, Huang D, Qu X, Zhu JF (2020) Effects of different harvesting ways on grain loss: based on the field survey of 3251 rural households in China. J Nat Resour 35:1043–1054. https://doi.org/10.31497/zrzyxb.20200503
Lu QA, Du XD (2020) The outsourcing choice of agricultural production tasks: implications for food security—a multiple-task based approach. In: The 2020 Agricultural & Applied Economics Association Annual Meeting. Kansas City, Missouri
Lu S, Cheng G, Li T et al (2022) Quantifying supply chain food loss in China with primary data: a large-scale, field-survey based analysis for staple food, vegetables, and fruits. Resour Conserv Recycl 177:106006. https://doi.org/10.1016/j.resconrec.2021.106006

Luo BL (2014) The theoretical trajectory and innovation direction of the agricultural management system: a case of Sichuan province. Reform 96–112

Mi Q, Li XD, Gao JZ (2020) How to improve the welfare of smallholders through agricultural production outsourcing: evidence from cotton farmers in Xinjiang. Northwest China. J Clean Prod 256:120636. https://doi.org/10.1016/j.jclepro.2020.120636

Nath B, Hossen M, Islam A et al (2016) Postharvest loss assessment of rice at selected areas of Gazipur district. Bangladesh Rice J 20:23–32. https://doi.org/10.3329/brj.v20i1.30626

Ndindeng SA, Candia A, Mapiemfu-Lamare D et al (2021) Valuation of rice postharvest losses in Sub-Saharan Africa and its mitigation strategies. Rice Sci 28:212–216. https://doi.org/10.1016/j.rsci.2021.04.001

Oguntade AE, Thylmann D, Deimling S (2014) Post-harvest losses of rice in Nigeria and their ecological footprint. Federal Ministry of Economic Cooperation and Development, Deutsche Gesellschaft fur Internatinale Zusammenarbeit (GIZ), Bonn, Germany

Parfitt J, Barthel M, MacNaughton S (2010) Food waste within food supply chains: quantification and potential for change to 2050. Philos Trans R Soc B Biol Sci 365:3065–3081. https://doi.org/10.1098/rstb.2010.0126

Picazo-Tadeo AJ, Reig-Martínez E (2006) Outsourcing and efficiency: the case of Spanish citrus farming. Agric Econ 35:213–222. https://doi.org/10.1111/j.1574-0862.2006.00154.x

Qu X, Kojima D, Nishihara Y et al (2021a) Can harvest outsourcing services reduce field harvest losses of rice in China? J Integr Agric 20:1396–1406. https://doi.org/10.1016/s2095-3119(20)63263-4

Qu X, Kojima D, Nishihara Y, et al (2021b) A study of rice harvest losses in China : Do mechanization and farming scale matter? Japanese J Agric Econ 23:83–88. https://doi.org/10.18480/jjae.23.0_83

Rosegrant MW, Magalhaes E, Valmonte-Santos RA, Mason-D'Croz D (2015) Returns to investment in reducing postharvest food losses and increasing agricultural productivity growth. Copenhagen Consensus Center, USA

Roy R (2013) Assessment of pre and post harvest losses in wheat and paddy crops in Uttar Pradesh. Agro-Economic Research Centre, University of Allahabad, Allahabad

Sadiya SS, Hassan II (2018) Postharvest loss in rice: Causes, stages, estimates and policy implications. Agric Res Technol Open Access J 15:111–114. https://doi.org/10.19080/artoaj.2018.15.555964

Sanneh L (2015) Effects of threshing and post-threshing recovery methods on postharvest losses in two varieties of rice. Kwame Nkrumah University, Kabwe, Zambia

Sarkar D, Datta V, Chattopadhyay KS (2013) Assessment of pre and post harvest losses in rice and wheat in West Bengal. Agro-Economic Research Centre, Visva-Bharati, Santiniketan

Sawicka B (2019) Post-harvest losses of agricultural produce. Sustain Dev 1:1–16. https://doi.org/10.1007/978-3-319-69626-3_40-1

Selvi R, Kalpana R, Rajendran P (2002) Pre and post harvest technologies to reduce yield losses in rice—a review. Agric Rev 23:252–261

Shafiee-Jood M, Cai X (2016) Reducing food loss and waste to enhance food security and environmental sustainability. Environ Sci Technol 50:8432–8443. https://doi.org/10.1021/acs.est.6b01993

Sheahan M, Barrett CB (2017) Review: Food loss and waste in Sub-Saharan Africa. Food Policy 70:1–12. https://doi.org/10.1016/j.foodpol.2017.03.012

Sivagnanam KJ (2013) Estimation of pre-and post-harvest losses in paddy crop in Tamil Nadu. Agro-Economic Research Centre, University of Madras, Chennai, India

Ssebaggala G, Kibwika P, Kyazze F, Karubanga G (2017) Farmers' perceptions of rice postharvest losses in Eastern Uganda. J Agric Ext 21:30–43. https://doi.org/10.4314/jae.v21i2.3

Stathers T, Holcroft D, Kitinoja L et al (2020) A scoping review of interventions for crop postharvest loss reduction in Sub-Saharan Africa and South Asia. Nat Sustain 3:821–835. https://doi.org/10.1038/s41893-020-00622-1

Taiwo A, Bart-Plange A (2016) Factors responsible for post-harvest losses and their effects on rice producing farmers: a case study of Afife and Aveyime rice projects in the Volta region of Ghana. Int Res J Eng Technol 3:1014–1022
Tang LQ, Liu Q, Yang WJ, Wang JY (2018) Do agricultural services contribute to cost saving? evidence from Chinese rice farmers. China Agric Econ Rev 10:323–337. https://doi.org/10.1108/CAER-06-2016-0082
Totobesola M, Delve R, d'Amour NJ et al (2022) A holistic approach to food loss reduction in Africa: food loss analysis, integrated capacity development and policy implications. Food Secur 14:1401–1415. https://doi.org/10.1007/s12571-021-01243-y
Veerangouda M, Sushilendra S, Prakash KV, Anantachar M (2010) Performance evaluation of tractor operated combine harvester. Karnataka J Agric Sci 23:282–285
Wang GM, Yi ZY, Chen C, Cao GQ (2016) Effect of harvesting date on loss component characteristics of rice mechanical harvested in rice and wheat rotation area. Trans Chinese Soc Agric Eng 32:36–42
Wolf CA (2003) Custom dairy heifer grower industry characteristics and contract terms. J Dairy Sci 86:3016–3022. https://doi.org/10.3168/jds.S0022-0302(03)73900-9
Wu LH, Hu QP, Wang JH, Zhu D (2017) Empirical analysis of the main factors influencing rice harvest losses based on sampling survey data of ten provinces in China. China Agric Econ Rev 9:287–302. https://doi.org/10.1108/CAER-03-2016-0036
Yang FT (2007) Analysis on the principal–agent relationship between agricultural machinery outsourcer and agricultural machinery operation service provider. Chinese Agric Mech, 15–18. https://doi.org/10.13733/j.jcam.issn.2095-5553.2007.01.004
Yi Q (2018) Adoption of agricultural mechanization services among maize farmers in China: impacts of population aging and off-farm employment. In: 30th International Conference of Agricultural Economists. Vancouver
Zhan YR (1995) Sampling survey and analysis of national grain post-harvest losses. Chian Grain Econ 4:44–47
Zhang XB, Yang J, Thomas R (2017) Mechanization outsourcing clusters and division of labor in Chinese agriculture. China Econ Rev 43:184–195. https://doi.org/10.1016/j.chieco.2017.01.012
Zorya S, Morgan N, Diaz Rios L (2011) Missing food: the case of postharvest grain losses in sub-Saharan Africa. The World Bank, Washington, DC

Chapter 3
Data Collection and Descriptive Statistics

In this chapter, the detailed description of the data used in this study is presented. After introducing two nationwide data used in this book, this chapter presents a detailed description of the estimation methods for the key variables and the statistical description of rice harvest losses.

3.1 Sample Design

The data used in this book comes from a large-scale survey of farm households named "Investigation and Evaluation of Rice Harvest Loss", which was organized by the research team of China Agricultural University (CAU) and the Research Centre for Rural Economy (RCRE) under the Ministry of Agriculture and Rural Affairs of China (MARA) from June to July 2016. The research team conducted this survey using the Rural Fixed Observation Point (RFOP) survey system of RCRE. The RFOP survey system was established in 1984 with the approval of the Secretariat of the Central Committee of the Communist Party of China and formally operated in 1986. It is a typical rural socio-economic survey system established to understand the production and operation of farm households. Currently, the RFOP survey system has covered 31 provinces (autonomous regions and municipalities) (hereinafter referred to as provinces), 357 counties (cities and districts), 360 villages, and over 23,000 farming and herding households, except for Hong Kong, Macao, and Taiwan, which has accumulated rich and informative survey data in the past 30 years (Qian 2019).

The sampling population covered 20,398 households in 335 fixed observation villages across 28 provinces, excluding Shanghai, Hainan, and Tibet. First, the top ten provinces for each crop (eight types of grain and oil varieties: wheat, rice, maize, soybean, rapeseed, peanut, potato, and sweet potato) were selected according to the crop production rankings in 2015. Considering the production and regional characteristics of each crop, the stratified sampling method was employed to choose two

X. Qu et al., *Harvest Loss in China*, The University of Tokyo Studies on Asia,
https://doi.org/10.1007/978-981-97-9156-9_3

counties in each province, two towns in each county, and two villages in each town. After performing aforementioned sampling for each crop, a total of 217 fixed observation point villages were selected. Then, within the selected sample villages, professional investigators from RFOP randomly selected 10 to 40 households according to the household distribution of each village. Finally, from the sampling population, a total of 4,170 households (covering all eight types of grain and oil varieties) were selected for a supplementary survey, which investigated food production and harvest losses.

In 2016, the research team conducted a pre-survey using questionnaires with 40 households in Hebei Province. Based on the issues identified in the pre-survey and feedback from experts, the research team improved the questionnaire and eventually formed a standardized structured questionnaire and research manual for data collection. To ensure the quality of the questionnaire, the research team distributed operation manuals to the professional investigators from RFOP and conducted centralized training for them twice at the end of April and the end of May in 2016. After the questionnaires were completed, the research team members reviewed the questionnaires to ensure the authenticity and reliability of the survey results to the greatest extent. Finally, a total of 3,739 questionnaires were collected (for all eight types of grain and oil varieties), with a total response rate of 89.66%.

Additionally, the RFOP office provided the corresponding data from their regular survey ("National Rural Fixed Observation Point Survey" in 2015) for use. Using "province code", "village code", and "household code" as the identification code (ID), the data obtained from the supplementary survey ("Investigation and Evaluation of Rice Harvest Loss") was merged with the regular survey data provided by the RFOP office. After excluding 249 questionnaires that could not be matched due to missing "village code" and "household code", a total of 3,490 valid matched questionnaires were obtained.

In 2015, the top ten rice-producing provinces[1] accounted for 81.80% of the total national production (left in Fig. 3.1). In the actual investigation, the same questionnaire was used for eight grain and oil varieties, and the survey was not conducted in their respective top ten provinces, but in a mixed sample of the eight grain and oil varieties. As a result, farmers growing rice are not only distributed in the top ten rice-producing provinces. Those farmers from outside the top ten rice-producing province are also taken into account. After excluding some missing data, there are 1,106 households. These 1,106 households covered 22 villages in 19 provinces,[2] accounting for 95.45% of China's rice production in 2015 (right in Fig. 3.1). In 2008, MARA (2008) issued the *Regional Layout Planning for Advantageous Agricultural Products* (Hereinafter referred to as *Planning*), which divided three advantageous regions for rice production: the Northeast Plain, the Yangtze River Basin, and the Southeast Coast (Table 3.1). The division of advantageous regions is based on

[1] The ten provinces are Hunan, Heilongjiang, Jiangxi, Hubei, Jiangsu, Anhui, Sichuan, Guangxi, Guangdong, and Jilin.

[2] Other nine provinces are Yunnan, Chongqing, Zhejiang, Guizhou, Fujian, Liaoning, Shandong, Shaanxi, and Tianjin.

regional resource endowments, comprehensively considering factors such as industrial foundations, market conditions, and ecological environment. This division can effectively represent the status of rice production and serve as key development areas for future rice cultivation.

Table 3.2 presents the sample provinces and their respective sample sizes. For some sample provinces that do not belong to any advantageous region, they are classified into the corresponding advantageous regions based on their natural resources, cultivation characteristics, and geographical location. The sample covers most of the areas in the three advantageous regions. The Yangtze River Basin covers more than half of the sample provinces, accounting for 66.55% of the total sample size. Samples distributed in the Northeast Plain and Southeast Coast account for 12.57% and 21.88%, respectively.

Fig. 3.1 Rice production share of top ten provinces and sample provinces in China (2015)

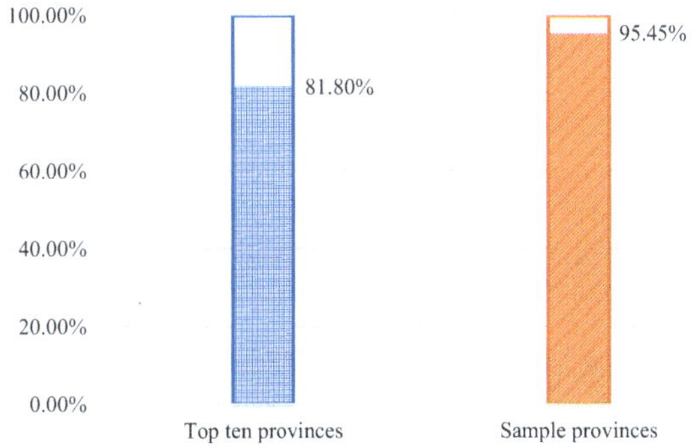

Table 3.1 Three advantageous regions for rice production

Regions	Location	Provinces
The Northeast Plain	Sanjiang Plain, Songnen Plain, and Liao River Plain	Liaoning, Jilin, Heilongjiang
The Yangtze River Basin	Sichuan Basin, the hilly and flat dam areas of the Yunnan-Kweichow Plateau, Dongting Lake Plain, Jianghan Plain, the southern part of Henan Province, Poyang Lake Plain, and the plains and hilly areas along Huaihe River and the Yangtze River	Jiangsu, Anhui, Jiangxi, Hubei, Hunan, Sichuan, Guizhou, Yunnan, Chongqing, Henan
The Southeast Coast	Hangzhou-Jiaxing-Huzhou Plain, Minjiang River Basin, Pearl River Delta, Chao-shan Plain, and the plains of Guangxi and Hainan provinces	Zhejiang, Fujian, Guangdong, Guangxi, Shanghai, Hainan

Data source Planning

Table 3.2 Sample distribution in three advantageous regions for rice production

Regions	Sample provinces	Sample
The Northeast Plain	Tianjin, **Liaoning**, **Jilin**, **Heilongjiang**, and Shandong	139 (12.57%)
The Yangtze River Basin	**Jiangsu**, **Anhui**, **Jiangxi**, **Hubei**, **Hunan**, **Sichuan**, **Guizhou**, **Yunnan**, **Chongqing**, and Shaanxi	725 (65.55%)
The Southeast Coast	**Zhejiang**, **Fujian**, **Guangdong**, and **Guangxi**	242 (21.88%)

3.2 Rice Harvest Loss Characteristics

3.2.1 Measurement of Rice Harvest Loss

Rice harvest losses measured in this book refer to the losses from the field to the household storage places, including reaping loss, threshing loss, winnowing loss, and field transport loss (see Sect. 1.3.1). In expressions of absolute value of loss, loss per unit area, and percentage loss, the percentage loss is chosen to measure rice harvest losses, which is widely used in literature (Sadiya and Hassan 2018). Additionally, using the loss rate facilitates comparisons of loss levels between different regions, harvesting methods, and farm scales. The rice harvest loss rate is expressed as the ratio of the total losses from these four stages to the sum of the total losses and harvested output, as shown in Eq. (3.1):

$$\begin{aligned}\mathrm{HLR} &= \frac{\text{harvest losses}}{\text{harvest losses } + \text{ PRO}} \\ &= \frac{L_{\mathrm{reap}} + L_{\mathrm{thr}} + L_{\mathrm{win}} + L_{\mathrm{tra}}}{\left(L_{\mathrm{reap}} + L_{thr} + L_{\mathrm{win}} + L_{\mathrm{tra}}\right) + \mathrm{PRO}} \times 100\%\end{aligned} \tag{3.1}$$

where HLR is the rice harvest loss rate. L_{reap} represents reaping loss, and it refers to scattering loss and uncut loss. L_{thr} is threshing loss, and it refers to grains remaining on the straw or scattered on the threshing floor when peeling the grain from the straw. L_{win} refers to winnowing loss and means grains blown away with impurities when separating grains from impurities; and L_{tra} represents field transport loss and refers to grains spilled on the road while being transported from the field to the storage places.[3] PRO denotes the final production quantity after harvesting.

Data collection of harvest losses is always a complicated work. Conducting field experiments to pick up and weigh rice that is left in each spot, such as the field, threshing floor, and road, may be a very reliable method. However, this approach is time-consuming, labor-intensive, and expensive, especially when the sample size is large and widespread. In this book, the loss data estimated by farmers themselves are used, which is the most frequently used method in the literature. Based on years of experience, farmers have fairly accurate understanding of the expected yield by the appearance of the harvest-ready crop. Although the credibility of farmers' self-reported losses may be biased, they can still accurately reflect the real situation in the case of large samples (Kaminski and Christiaensen 2014). furthermore, in the absence of any experimental evidence to compare the loss rates based on self-reports and direct measurement, farmers' estimations can reasonably be considered reliable (Sheahan and Barrett 2017). Therefore, professional investigators from RFOP asked the farmers to recall and estimate the losses at these four stages. When using combine harvesters, farmers usually estimated the total losses during the reaping, threshing, and winnowing stages.

---

[3] Please refer to Chap. 1 for the specific causes of these losses.

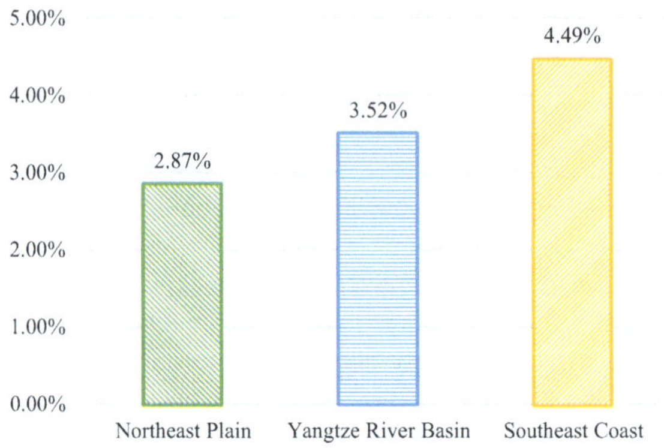

Fig. 3.2 Average HLR of rice in three advantageous regions

3.2.2 *Rice Harvest Loss Characteristics*

Overall, in 2015, the average rice harvest loss rate in China was 3.65%. Figure 3.2 shows the average rice harvest loss rates in three advantageous regions. This figure falls within the range of 1.23–5.5% for rice harvest loss rate in China (see Table 2.1 in Sect. 2.1.2). It means that nearly 8 million tons of rice and more than 1 million hectares of farmland were wasted. Moreover, the Northeast Plain had the lowest average rice harvest loss rate, at only 2.87%. The Northeast Plain is located among the three plains in China, and its flat topographic advantage may be the reason for the low harvest loss rate. The average harvest loss rate of rice in the Southeast Coast was the highest among the three advantageous regions, reaching 4.49%. As a coastal region, the harvesting season is often affected by typhoons and heavy rainfall, resulting in higher rate of lodging and thus greater harvest losses (Huang et al. 2018). The average harvest loss rate of rice in the Yangtze River Basin was 3.52%. This region is composed by plains, basins, and hills, making its average harvest loss rate higher than that in the Northeast Plain and lower than that in the Southeast Coast.

Figure 3.3 displays the average harvest loss rates for different farming scales. Based on the deciles of rice planting area, 1106 households were divided into 10 groups. Generally, the average rice harvest loss rates decreased as the farming scale expands. As the planting area increases, profitability becomes increasingly important. Larger farms are more likely to invest in agricultural technologies, such as pest and disease control, thereby reducing harvest losses. Additionally, they have more access to machinery, such as combine harvesters and harvest outsourcing services (Zhou 2017).

Table 3.3 gives the average harvest loss rates of rice at each stage. In segmented harvesting, the loss rate during the reaping stage was higher than those in other harvesting stages, accounting for more than 60% of the total harvest loss rate. This was followed by the threshing stage, which accounted for nearly 20%. In both segmented and combine harvesting, the loss rate during the field transport stage was the smallest among the four operations.

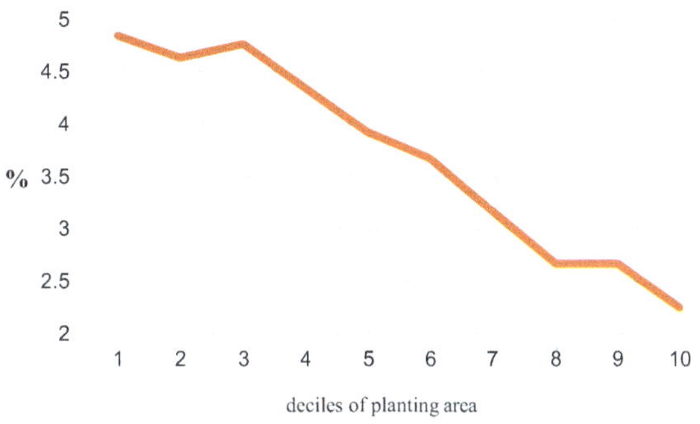

Fig. 3.3 Average HLR of rice on different farming scales

Table 3.3 Average HLR of rice for each stage

	Segmented harvesting (%)	Combine harvesting (%)
Reaping	2.48	3.27
Threshing	0.76	
Winnowing	0.42	
Field transport	0.22	0.12

Note (1) There is only one total ratio in combine harvesting because combine harvesters combine the first three stages into one process
Data source Author's calculation based on the survey

Table 3.4 presents farmers' choice of harvest methods and the corresponding harvest loss rates. There was little difference between the adoption of combine harvesting and segmented harvesting, with a slightly higher adoption of segmented harvesting. A total of 595 farmers adopted segmented harvesting, accounting for 53.80% of the sample. This implies that there is still room for the promotion of combine harvesters. The average rice harvest loss rate for segmented harvesting was 3.88%, which was higher than the 3.39% for combine harvesting. Compared to combine harvesting, segmented harvesting involves more stages, and each stage inevitably incurs losses during the operation (Wu et al. 2017). Meanwhile, combine harvesters can quickly complete the harvesting process by combining the reaping and threshing stages, thereby avoiding losses caused by adverse weather conditions.

To further understand the rice harvest loss rates with different harvest methods, Fig. 3.4 illustrates the rice harvest loss rates in three advantageous regions. Interestingly, the harvest methods that resulted in higher losses varied across different

Table 3.4 Farmers' choice about harvest methods and their average HLR of rice

Harvest methods	Sample	HLR (%)
Segmented harvesting	595 (53.80%)	3.88
Combine harvesting	511 (46.20%)	3.39

Data source Author's calculation based on the survey

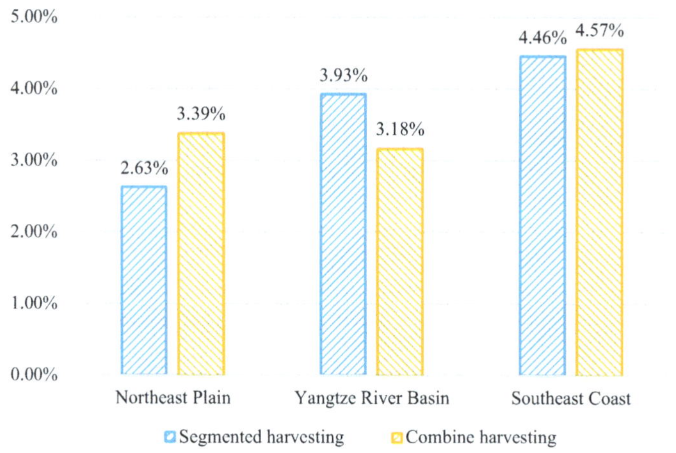

Fig. 3.4 Average HLR rice for different harvest methods in three advantageous regions

regions. Segmented harvesting did not always result in higher losses. In the Northeast Plain, the average loss rate of rice harvested by combine harvesters was 3.39%, higher than the 2.63% for segmented harvesting. The Northeast Plain is characterized by flat terrain, which should be more conducive to mechanical operations. However, why does the combine harvesting have a higher loss rate? The reason could be the moral hazard mentioned earlier. While flat terrain facilities the operation of combine harvesters, it also creates conditions for moral hazard. As mentioned in Chap. 1, one of the manifestations of moral hazards could be accelerated harvesting, which requires favorable terrain conditions. It is difficult to accelerate forward speed of machines on rugged and narrow farmland. The relatively wide and flat terrain in the Northeast Plain creates favorable conditions for rapid mechanical harvesting. Moreover, the large average planting areas in the Northeast Plain pose challenges for supervision and management.

In contrast, in the Yangtze River Basin, the average rice harvest loss rate for combine harvesting was 3.18%, which was smaller than the 3.93% for segmented harvesting. The Yangtze River Basin has more hills and a relatively smaller average planting area, making it difficult for combine operators to speed up harvesting. At this point, the advantages of combine harvesters are more evident, e.g., the quick harvesting (compared to segmented harvesting) can help avoid losses caused by delay harvesting.[4]

Table 3.5 presents the average harvest loss rates of rice for different services. Most of the harvesting operations were carried out by outsourcing services. Specifically, 651 farmers used harvest outsourcing services, accounting for nearly three-fifths of the total sample. Once again, this demonstrates the importance of studying the effect of harvest outsourcing services on rice harvest losses. The average harvest loss

[4] This of course does not mean that there is no moral hazard, as this is only a preliminary statistic before controlling for other possible influencing factors.

Table 3.5 Average HLR of rice for different services

Service	Sample	*HLR*
Self-service	455 (41.14%)	3.87%
Outsourcing service	651 (58.86%)	3.50%

Note 1) Self-service represents farmers who did not purchase harvest outsourcing services
Data source Author's calculation based on the survey

rate of rice for outsourcing services was 3.50%, slightly lower than the 3.87% for self-service. This may be attributed to the fact that the vast majority of outsourcing service providers used combine harvesters.

Figure 3.5 shows the average rice harvest loss rates for different services in three advantageous regions. In the Northeast Plain, the average rice harvest loss rate for outsourcing services was 3.67%, much higher than the 2.39% for self-service. This aligns with the analysis in Fig. 3.4, implies moral hazard in the work of service providers. Hence, outsourcing services caused higher harvest loss rates compared to self-service. In the Yangtze River Basin, the average harvest loss rate for outsourcing services was 3.27%, lower than that of self-service (4.05%). This could be attributed to certain moral hazard behaviors by service providers, such as increasing forward speed, constrained by relatively rugged and narrow farmland. Service providers have a higher adoption of combine harvesters, which can harvest in a timely manner and mitigate greater losses from exposure to severe weather. Additionally, service providers may possess more advanced machinery and professional skills.

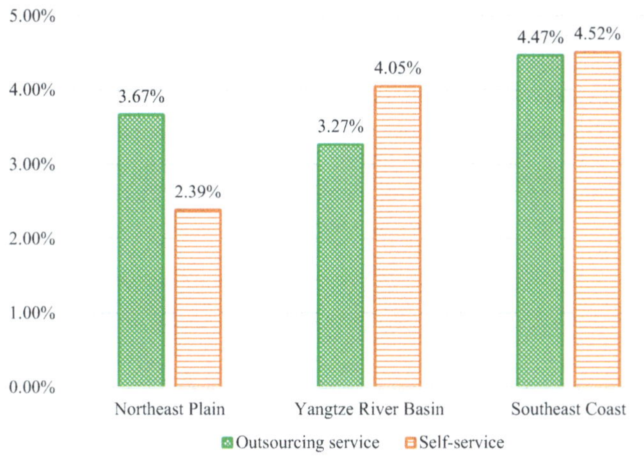

Fig. 3.5 Average HLR of rice for different services in three advantageous regions

3.3 Summary

This chapter introduces two national farm household surveys that constitute the data for this study. Using data from these surveys, which includes 1,106 farm households across 19 provinces in China, the current status and characteristics of rice harvest losses in China are described.

Rice harvest losses are measured by the loss rate, which is the ratio of total losses during the stages of reaping, threshing, winnowing, and field transport stages to the total production. Overall, the average rice harvest loss rate in China was 3.65%, which is generally consistent with other studies. This means that approximately 8 million tons of rice and more than 1 million hectares of farmland were wasted.

The rice harvest loss rates varied by regions and farm scales. The Northeast Plain had the lowest average rice harvest loss rate at 2.87%, while the Southeast Coast had the highest at 4.49%. The average rice harvest loss rate in the Yangtze River Basin was 3.52%, which falls between the other two regions. As farm scales increased, the average rice harvest loss rates tended to decrease.

The average rice harvest loss rate for combine harvesting was lower than that for segmented harvesting. Less than half of the farmers used combine harvesters, with an average rice harvest loss rate of 3.39%, compared to 3.88% for segmented harvesting.). For segmented harvesting, the reaping stage had the highest loss rate among the four stages. However, combine harvesting in the Northeast Plain caused greater losses than segmented harvesting. This may be due to the relatively flat terrain and large farmlands, which present conditions for moral hazard and difficulties for monitoring and management.

The average rice harvest loss rate of service providers was lower than that of farmers. More than half of the farmers purchased harvest outsourcing services. The average rice harvest loss rate by service providers was 3.50%, which was lower than 3.87% by farmers. This is likely because the majority of service providers used combine harvesters and they had more advanced machines and more specialized skills. However, in the Northeast Plain, rice harvest loss rate by service providers was higher than that by farmers. This coincides with the moral hazard discussed earlier regarding harvest outsourcing services.

References

Huang D, Yao L, Wu LP, Zhu X Di (2018) Measuring rice loss during harvest in China: based on experiment and survey in five provinces. J Nat Resour 33:1427–1438. https://doi.org/10.31497/zrzyxb.20170810

Kaminski J, Christiaensen L (2014) Post harvest loss in Sub Saharan Africa—what do farmers say? Glob Food Sec 3:149–158. https://doi.org/10.1016/j.gfs.2014.10.002

MARA (2008) Regional layout planning of national superior agricultural products. http://www.moa.gov.cn/nybgb/2008/djiuq/201806/t20180611_6151652.htm. Accessed 17 Jan 2022

Qian WR (2019) Societal development in rural China. Zhejiang University Press, Zhejiang, China

Sadiya SS, Hassan II (2018) Postharvest loss in rice: Causes, stages, estimates and policy implications. Agric Res Technol Open Access J 15:111–114. https://doi.org/10.19080/artoaj.2018.15.555964

Sheahan M, Barrett CB (2017) Review: food loss and waste in Sub-Saharan Africa. Food Policy 70:1–12. https://doi.org/10.1016/j.foodpol.2017.03.012

Wu LH, Hu QP, Wang JH, Zhu D (2017) Empirical analysis of the main factors influencing rice harvest losses based on sampling survey data of ten provinces in China. China Agric Econ Rev 9:287–302. https://doi.org/10.1108/CAER-03-2016-0036

Zhou J (2017) The dual division of agricultural socialization service and its consequences under the background of land circulation. J Nanjing Agric Univ (Social Sci Ed) 17:141–151

Chapter 4
The Moral Hazard in Harvest Outsourcing Service

This chapter will examine the moral hazard in harvest outsourcing services. In order to study the reduced effort by service providers, work attitudes toward harvesting work will be compared between service providers and farmers. In this chapter, the presence of moral hazard among service providers will be studied in terms of farming scale and part-time farming.

4.1 Introduction

For several years, agricultural mechanization has been a key goal of agricultural development in China. However, the decentralized management of smallholders in China poses challenges for the application of agricultural machinery (Zhang et al. 2018). Small-scale farmers are reluctant to invest in specialized agricultural machinery (Ruttan 2000). In Southeast Asian countries, only by integrating land into large contiguous fields can the adoption of combine harvesters improve the prospects of rice harvesting (Pingali 2007). Investment in agricultural machinery will benefit farms until their average farm size increases to 3 hectares or larger (Otsuka 2013).

Since the enactment of the *Rural Land Contracting Law* in 2003, China has introduced a series of measures to encourage land transfer, including the establishment of land transfer service centers and the confirmation of land rights. However, the average farm scale has hardly changed (Ji et al. 2016). Efforts to achieve scale management through land consolidation to create favorable conditions for the use of machinery have progressed slowly (Sheng et al. 2017; Zhang et al. 2018). Nevertheless, China has made considerable progress in agricultural mechanization (Zhang et al. 2017); from 1980 to 2019, the number of combine harvesters in China increased from 0.03 millions to 2.13 millions, a 70-fold increase (NBSC 2019).

Agricultural machinery outsourcing services have played a crucial role in China's agricultural mechanization (Sheng et al. 2017). Although small-scale

X. Qu et al., *Harvest Loss in China*, The University of Tokyo Studies on Asia,
https://doi.org/10.1007/978-981-97-9156-9_4

farmers cannot afford agricultural machinery, they can afford much cheaper harvest outsourcing services. Thus, whether individuals or organizations, purchasing agricultural machinery to respond to the demand for mechanical outsourcing services is profitable.

Many studies have examined agricultural machinery outsourcing services (Yang et al. 2013), such as their impact on agricultural production (Deng et al. 2020) and farmers' willingness to outsource services (Ji et al. 2017). However, few studies have explored the relationship between farmers and service providers. Given the prevalent principal–agent issues in outsourcing services in the secondary and tertiary sectors, the possibility of principal–agent issues in agricultural outsourcing services cannot be ignored (Lu and Du 2020). Harvest outsourcing service is a package rent out of agricultural machinery and machine operators (Belton et al. 2018). It enables the division of labor between farmers and mechanization service providers (Zhang et al. 2017). Service providers, as hired labor force different from family labor force, are not as responsible as those who work on their own farmland (Coelli and Battese 1996). Farmers entrust harvesting task, originally completed by household members, to other individuals or organizations with agricultural machinery, paying corresponding service fees to them. When farmers outsource harvesting services, they act as principals, while the service providers act as agents. On the one hand, service providers and farmers have different goals. Service providers offer harvesting services in return for service fee, which is based on the area serviced. The larger the serviced area, the higher the service fee. Farmers aim for efficient harvesting with minimal harvest loss. On the other hand, service providers and farmers have asymmetric information. Due to the non-standardized outsourcing service market and the lack of written contracts (Cai and Liu 2019), farmers are the less informed party in regard to the agents' skills and the maintenance of machines. Therefore, due to inconsistent goals and information asymmetry, service providers, as rational economic decision makers, may use their information advantage to maximize their self-utility through lower levels of effort (such as, increasing forward speed), thereby harming the interests of farmers (Pandey et al. 2013). In addition, the natural characteristics of agricultural production, such as long production cycles and susceptibility to natural environment, make agricultural supervision difficult to implement, increasing the potential for moral hazards (Cai and Liu 2019). Several studies have theoretically analyzed and discussed the possible moral hazards in agricultural outsourcing services. Based on the game theory and principal–agent theory, Cai and Liu (2019) and Huan and Hou (2020) argued that service providers may implement extensive operations and reduce service quality to pursue their profit maximization.

In brief, there is a moral hazard in harvest outsourcing services, which means that service providers may reduce their effort levels for profit maximization when providing harvesting services. This chapter will provide one of the few empirical analyses of service providers' reduced effort level in harvest outsourcing services. Considering that faming scale and part-time farming may change farmers' effort levels toward farming, moral hazard in outsourcing services is studied in sub-samples divided according to farming scale and part-time farming degree.

4.1.1 Faming Scale

Farms of different planting scales have widely differences in resource endowments. Large-scale farms are primarily commercial and have a higher demand for agricultural technology (Zhang and Qian 2008; Hu et al. 2022), which makes them more risk-averse (Wilk et al. 2013). Small-scale farms often lack access to decent inputs (Murphy 2012), and their goal is primarily to meet their own food demand. Although outsourcing services make mechanical harvesting available to farms of all scales, service providers may exert higher effort levels when serving large-scale farms because the large and concentrated farmland of large-scale farms are attractive service targets to them (Zhou 2017; Li et al. 2023). However, as farm scale increases, farm management may become more challenging, making supervision difficult. As analyzed in Chap. 3, large farmland may make supervision and management difficult and give chances for moral hazard behaviors, such as increased forward speed of machines. Therefore, when studying moral hazards by comparing effort levels between service providers and farmers, it is necessary to consider farming scales.

4.1.2 Part-Time Farming

Part-time farming can lead to varying effort levels toward farming among different farmers. By 2020, the total number of rural migrant workers nationwide has reached 285.6 million (NBSC 2021). A large number of rural labors, especially those under 40 years old, are engaged in non-farm work (Li et al. 2013a). Part-time farming is the allocation of household labors between farm and non-farm activities (Xu et al. 2019) or the division of time between farm and non-farm activities by household labors (Kimhi 2000). Therefore, part-time farming inevitably reduces the household agricultural labor force or farming time, thereby altering farmers' effort level toward farming. In Germany (Pfeffer 1989) and Australia (Weiss 1999), part-time farmers have low expectations for continuing farming in the future. The extensive participation of rural labor in non-farm employment will lead to a decline in both the quantity and quality of labor engaged in agriculture (Hao et al. 2013). Increased non-farm income is not invested in agricultural production but rather in housing and durable consumer goods (De Brauw and Rozelle 2008). Azam and Gubert (2006) argued that remittances from non-farm labors would lead to free-rider problem for farm labors. The production enthusiasm of farm labors would be weakened, causing their low efficiency. In China, land is extensively managed by farm labors (Hao et al. 2013) through reduced cropping index (Li et al. 2013b) and abandonment (Xu et al. 2019).

Some studies thought that part-time faming is not a way out of agriculture (Kimhi 2000). The exit rates of faming decrease with part-time farming (Kimhi 2000; Breustedt and Glauben 2007). In the case of harvest outsourcing services, the demand for harvest outsourcing services is likely to be higher on part-time farms, because they allocate more household labors to non-farm sectors. Therefore, it is also necessary to take into account part-time farming when studying moral hazards through comparison of effort level of service providers and farmers.

4.2 Data and Method

4.2.1 Data and Variable

The data used in this chapter is from the dataset presented in Chap. 3.[1] It covers 1106 rice farmers across 19 provinces in China, recording their rice harvest losses and production information in 2015. This section focuses on the variables used in this chapter.

4.2.1.1 Dependent Variable

To study the moral hazard in harvest outsourcing services, it is first necessary to measure the moral hazard. Moral hazard is an abstract concept. Therefore, we need to find a proxy variable for it. Moral hazard behaviors can manifest as increasing the forward speed of machines, ignoring rice in the corner, and neglecting adjustments of machine settings for rice growing conditions. These actions are difficult to obtain from farm household surveys. Field investigation could be a good way to observe these behaviors. However, the on-site investigation itself may also influence operators' behaviors, resulting in unobservable of these behaviors when the investigation is informed.

As mentioned in Chap. 1, the moral hazard of service providers is reflected in the fact that they reduce their own effort level in providing harvesting services. However, the original effort intensity of service provider cannot be directly observed. As Adam Smith noted, the agent's efforts are inferior to those of the principals (Smith 1937). Therefore, using the effort level of farmers as a criterion, service providers' reduced effort level will be studied by comparing the effort level of service providers and farmers. Specifically, if the effort level of service providers is lower than that of farmers, they are considered to have reduced effort level, indicating moral hazard. Conversely, if the effort level of service providers is not lower than that of farmers, then they have no reduced effort level, indicating no moral hazard.

In this study, work attitude is used as the proxy variable for effort level. Then, the presence of moral hazards is determined through the comparison of the work attitudes of service providers and farmers. If service providers' work attitude is less serious than that of farmers, service providers are considered to have reduced effort

[1] Please refer to Chap. 3 for more information about the dataset.

level. Conversely, if service providers' work attitude is not less serious than that of farmers, service providers are considered to have no reduced effort level.

In the questionnaire, there are three-point scale (fine, general, rough) for "operator's work attitude when reaping". Here, the operator can be either the service provider or the farmer. Specifically, if the farmer used harvest outsourcing services, the operator was the service provider. If the farmer did not use harvest outsourcing services, the operator was the farmer. The rating of the operator's work attitude is determined by farmers based on their observations during harvesting process, such as harvest losses, forward speed of machines, and other actions of operators they observed. Farmers are the most knowledgeable about harvesting, thus, they can evaluate both service providers' and their own work attitudes. Dissatisfaction among farmers with harvest outsourcing services validates this (Cai and Liu 2019). Although work attitude is a subjective variable, in the absence of objective variables, a reliably estimated subjective variable is an acceptable approach to capture abstract concepts. Similar subjective evaluations are frequently used in sociological studies, such as happiness and satisfaction (Diener et al. 1999). We believe that these estimations are reliable for large samples.

For simplicity, we categorized work attitude into two levels: "serious" and "not serious". When farmers rated it as "fine", it means that operator's work attitude was serious. When farmer rated it as "general" or "rough", it means that operator's work attitude was not serious. The dummy variable "WA" is used to denote work attitude. It means that "is service provider or farmer serious about harvesting?"

$$\mathrm{WA} = \begin{cases} 0, & \text{if service provider/farmer's work attitude was not serious} \\ 1, & \text{if service provider/farmer's work attitude was serious} \end{cases} \tag{4.1}$$

4.2.1.2 Core Independent Variable

By studying the effect of harvest outsourcing services on operators' work attitudes, comparisons between service providers and farmers' work attitudes were made, suggesting that the core independent variable is the purchase of harvest outsourcing services.

Typically, harvest outsourcing services are involved in the first two harvesting operations, namely reaping and threshing. In combine harvesting, if the combine harvester was supplied and operated by a service provider, it signifies that the farmer purchased harvest outsourcing services. In segmented harvesting, if both reaping and threshing were undertaken by service providers using their machines, then the farmer was considered to have purchased harvest outsourcing services.

The dummy variable "*Ser*" is used to denote harvest outsourcing services, which means "did the farmer use harvest outsourcing services?" Thus, OS = 1 means that the farmer used harvest outsourcing services; OS = 0 means that the farmer did not use harvest outsourcing services.

$$\mathrm{OS} = \begin{cases} 0, & \text{if farmers did not use harvest outsourcing services} \\ 1, & \text{if farmers used harvest outsourcing services} \end{cases} \tag{4.2}$$

4.2.1.3 Covariates

To ensure the accuracy of estimations, three types of control variables that may affect operators' work attitudes toward harvesting work are include: (1) production and harvesting conditions, (2) household and individual characteristics, and (3) regional control variables.

Production and Harvesting Conditions

Production and harvesting conditions that may affect operators' work attitudes include harvest methods (Com), weather conditions (Wea), pest and disease conditions (Pest), rice planting area (Area), land terrain (Flat), distance from the homestead to the nearest paved road (Htor), labor shortage (Labor), farmers' attitude toward harvest losses (Sav), and sale price of rice (Price).

Harvest methods (Com) Different harvest methods imply different service fees, which may affect operators' work attitudes. There are two rice harvest methods in China, namely combine harvesting and segmented harvesting (Qu et al. 2021a) (or direct harvesting and indirect harvesting (Alizadeh and Allameh 2013)). Briefly, combine harvesting refers to the use of combine harvesters and the rest are segmented harvesting. Combine harvesters integrate the first three harvesting operations—reaping, threshing, and winnowing—into a single process. Although combine harvesters can significantly reduce the labor force engaged in harvesting and shorten the harvesting period, they are large in size and costly. Therefore, in economically disadvantaged or geographically complex regions, farmers prefer segmented harvesting. Segmented harvesting, as the name implies, means that each harvesting operation is completed separately.[2] Therefore, the service fee for combine harvesting is higher than that for segmented harvesting (Poungchompu and Chantanop 2016). The dummy variable "Com" is used to represent harvest methods ("did farmers use combine harvesting?"), where Com = 1 indicates the use of combine harvesting, while Com = 0 indicates the use of segmented harvesting.

Weather conditions (Wea) If bad weather happens during harvesting process, operators may want to finish harvesting as soon as possible, which could lead to rough work. Meanwhile, bad weather itself increases harvest losses, making it difficult to determine whether the increased losses are due to bad weather or moral hazard,

[2] Segmented harvesting contains four types, based on whether a machine is used in the reaping and threshing stages: A) hand reaping and hand threshing; B) machine reaping and hand threshing; C) hand reaping and machine threshing; D) machine reaping and machine threshing, carried out in separate stages.

thereby increasing the likelihood of service providers' reduced effort level. Investigators asked farmers to recall the weather types during harvesting, including normal weather, heavy rain, strong winds, etc.[3] To simplify the analysis, weather condition is represented as a 0–1 dummy variable. If the weather type is normal weather, Wea = 0, otherwise Wea = 1.

Pest and disease conditions (Pest) Similar to weather conditions, pest and disease increase the likelihood of inefficient work because it is difficult to distinguish between harvest losses caused by pest and disease and those caused by moral hazard. Investigators inquired about farmers' estimation of pest and disease condition, which contained three options: no pest, slight or general pests, and sever pests. If the answer is "no pest", then Pest = 1. If the answer is slight or general and sever pests, Pest equals 2 and 3, respectively.

Rice planting area (Area) The larger the planting area, the more important the rice production is to farmers, and the higher the service fees for service providers. Therefore, the operator is more likely to work seriously.

Land terrain (Flat) Rugged farmland increases the difficulty of harvesting, especially for mechanical operations. This may negatively affect operators' work attitudes. There are four options in the questionnaire to describe the terrain: flat, sloped, depressed, and other. Dummy variable Flat is assigned the value of 1 if farmers reported the farmland as flat; otherwise, it is 0.

Distance from the homestead to the nearest paved road (Htor) The farther the distance from homestead to the nearest paved road, the longer the dirt road to reach the farmland. Driving on dirt roads requires extra caution. Additional, farmers who are economically disadvantaged and elderly are more likely to reside far from paved roads (Fan et al. 2000; Dercon et al. 2009). They typically dislike losses and are likely to take harvesting seriously, which could also affect the work attitude of service providers who serve them.

Labor shortage (Labor) One method to reduce the probability of moral hazards is through supervision. Labor shortages create opportunities for service providers and farmers to perform inefficient or inattentive work. Dummy variable Labor takes the value of 1 if farmers reported a shortage of labor when harvesting; otherwise, it is 0.

Farmers' attitude toward harvest losses (Sav) Farmers' attitudes toward harvest losses can affect operators' work attitudes. Farmers who are averse to harvest losses will be more careful when harvesting or more likely to monitor service providers, thus reducing the possibility of service providers' laziness. Whether farmers picked up leftover rice is used to capture farmers' attitudes toward harvest losses. If farmers gathered the rice left in the field after harvesting, they are deemed to have a sense of food saving, and dummy variable Sav is set to 1; otherwise, it is set to 0.

---

[3] The questionnaire contains the types of weather only, without a specific indicator of heavy rain or strong wind. It is based on farmers' observation during harvesting.

Sale price of rice (Price) The increase in sale price of rice may make farmers more dedicated to harvesting. However, for service providers, the rise in sale price of rice may also increase farmers' demand for harvesting services. Providing services to more farmers may negatively affects service providers' work attitudes. Specifically, Price refers to the sale price of rice in the first three months after harvesting. If no rice was sold in the first three months, the sale price in the first six months after harvesting is used, and so on. The missing values are filled with the average sale price of the village in the first three months. Although the sale price is formed after harvesting, we believe that farmers have enough rationality and knowledge to predict future price with reasonable accuracy (Lucas 1967).

Household and individual characteristics

Household and individual characteristics that may affect operators' work attitudes include the gender (Gen), age (Age), education years (Edu), and agricultural training experience (Train) of the household head, total family income (Tinc), and the proportion of rice income (Rincs). Table 4.2 summarizes their definitions and measurements.

Regional control variables

In addition to the above control variables, three advantageous regions of rice are introduced to avoid the effect of unobservable regional variables. These three regions are the Yangtze River Basin, the Northeast Plain, and the Southeast Coast issued by the MARA of China (MARA 2008). As we introduce in Chap. 3, this division takes into account regional resource endowments, industrial foundations, market conditions, and ecological environments, effectively reflecting the current status of rice production and outsourcing service market in China.

4.2.2 Logit Model

Let's consider a dummy response variable y (0 = failure and 1 = success) with respect to the outcome of the explanatory variable $x_1, \ldots, x_k$. Let $P(\mathbf{x})$ represents the probability of success when the random variable y is 1. The logistic regression model is given by

$$P(\mathbf{x}) = \frac{\exp(\beta_0 + \beta_1 x + \ldots + \beta_k x_k)}{1 + \exp(\beta_0 + \beta_1 x + \ldots + \beta_k x_k)} \tag{4.3}$$

where $\beta_0, \ldots, \beta_k$ are parameters to be estimated. $\mathbf{x}$ is the vector of all explanatory variables $x_1, \ldots, x_k$. This model is represented by a nonlinear, monotonous, S-shaped function with codomain [0, 1].

The odds of success are defined as the ratio between the probability of success and the probability of failure and are represented by

$$\text{Odds}(x) = \frac{P(x)}{1 - P(x)} = \exp(\beta_0 + \beta_1 x + \ldots + \beta_k x_k) \tag{4.4}$$

Note that the logarithm of the odds is a linear function of the explanatory variable $x_1, \ldots, x_k$. The logarithm of the odds is known as the Logit and is represented by

$$\text{logit}[P(x)] = \ln\left(\frac{P(x)}{1 - P(x)}\right) = \beta_0 + \beta_1 x + \ldots + \beta_k x_k \tag{4.5}$$

The transformation (4.5) in the logistic models means that the Logit model possesses certain significant properties of the simple linear regression model: $\text{logit}[P(x)]$ has linear parameters; it may be continuous or vary from $-\infty$ to $+\infty$.

In this study, we establish a multivariate Logit regression model to study the effect of harvest outsourcing services on operators' work attitudes (0 = not serious work attitude and 1 = serious work attitude):

$$\ln\left(\frac{\text{Prob(WA} = 1)}{\text{Prob(WA} = 0)}\right) = \alpha_0 + \alpha_1(\text{OS}) + \sum_{i=2}^{k} \alpha_i c_i \tag{4.6}$$

Prob(WA = 1) is the probability that work attitude equals 1, while Prob(WA = 0) is the probability that work attitude equals 0. The probability of WA = 1 to the probability of WA = 0 (Prob(WA = 1)/Prob(WA = 0)) is called odds. Therefore, the dependent variable is the natural logarithm of the odds (log odds). OS is a dummy variable that equals 1 if farmers purchased harvest outsourcing services; otherwise, 0. $c_2 \ldots c_i$ are covariates that have impacts on operators' work attitudes, which are listed in "Covariates". α_0 is the intercept. The regression coefficients $\alpha_1, \alpha_2, \ldots, \alpha_i$ are given in units of log odds, which indicate the amount of change expected in the log odds when there is a one-unit change in the predictor variable with all of the other variables in the model held constant.

As we mentioned above, there are two harvest methods. To study if the effect of outsourcing services on work attitudes is affected by the harvest methods. We add the cross term of harvest outsourcing services and harvest methods to get the Eq. (4.7):

$$\ln\left(\frac{\text{Prob(WA} = 1)}{\text{Prob(WA} = 0)}\right) = \beta_0 + \beta_1(\text{OS}) + \beta_2(\text{Com}) + \beta_3(\text{Ser} \times \text{Com}) + \sum_{i=4}^{k} \beta_i z_i \tag{4.7}$$

Com is harvesting method, which takes on 1 if the farmer used combine harvesting and 0 if the farmer used segmented harvesting. $z_4 \ldots z_i$ are covariates. β_0 is the intercept, while $\beta_1, \beta_2, \beta_3 \ldots \beta_i$ are estimated coefficients.

Since the cross term of harvest outsourcing services and harvest methods is added, the base/default group is farmers who used segmented harvesting and did not purchase harvest outsourcing services. Then the marginal effect of OS (β_1) indicates the difference of serious attitudes probability between outsourcing services using segmented harvesting and self-service using segmented harvesting. The marginal effect of Com (β_2) indicates the difference of farmers' serious attitudes probability between combine harvesting and segmented harvesting. To compare the difference of serious attitudes probability between harvest outsourcing services using combine harvesting and self-service using combine harvesting and the difference of service attitudes probability between combine harvesting using harvest outsourcing services and segmented harvesting using harvest outsourcing services, Stata command "Lincom" will be used to calculate the difference and assess statistical significance. The Logit procedure used for obtaining the marginal effect estimates are carried out using STATA 15.

4.2.3 *Classification of Farm Types*

4.2.3.1 Farming Scale

In China, there is no unified classification for farm scale. Existing literature employs roughly three methods to categorize farm scale. The first method is to classify the farm scale based on statistical characteristics of sample data. Li et al. (2019) classified farms into small-scale farms (area $\leq$ 0.25 ha), medium-scale farms (0.25 ha $<$ area $\leq$ 0.6 ha), and large-scale farms (area $>$ 0.6 ha) based on the planting area of the sample. The second method is to classify the farm scale based on the criteria of family farm or large grain production household, mainly found in the studies of new agricultural business entities. Chen and Tang (2020) classified the farms into small-scale farms (area $<$ 2 ha) and large-scale farms (area $\geq$ 3.33 ha). Liu (2021) regarded those with an area of less than 3.33 ha as small-scale farms and those larger than 3.33 ha as family farms. The third method does not classify farm scale directly, but only examines the relative size of farms. Huang and Luo (2020) categorized farms into different scales such as 0.33 ha, 0.67 ha, 1 ha, and so on.

Since this study does not discuss what size different farms should be, it simply analyzes the effect of outsourcing services on work attitudes across different farm scales. Therefore, farm scale here is a relative concept. Referring to the first method, this study uses the statistical characteristics of the sample, specifically the median of rice planting area (0.22 ha), to divide the sample into small-scale farms and large-scale farms. This criterion closely aligns with that used by Li et al. (2019). Specifically, farms with rice planting area less than 0.22 ha are classified as small-scale farms, while farms with rice planting area greater than or equal to 0.22 ha are classified as large-scale farms.[4]

[4] Samples that used manual reaping and manual threshing were not included when counting large-scale farms, as large-scale farms are unlikely to adopt these methods.

Table 4.1 Definition of part-time farm and business farm

Classification	Definition	Farm income/total income (%)
Part-time farm	Farm income/total income ≤ 1/2	24.48
Business farm	Farm income/total income > 1/2	77.20

Note In this study, the proportion of rural household management income is used to represent the proportion of farm income
Data source Author's calculation based on the survey

4.2.3.2 Part-Time Farming

In existing literature, the proportion of farm income or the number of days engaged in non-farm employment is often used to define different degrees of part-time farming. In this study, farms are categorized into part-time farms and business farms based on the proportion of farm income.

Table 4.1 gives the definitions of part-time farms and business farms. Part-time farms refer to those where the proportion of farm income to total household income is less than or equal to 50%, while business farms refer to those where farm income constitutes more than 50% of total household income. In our sample, the average proportion of farm income to total household income was 24.48% for part-time farms, while for business farms, the average proportion of farm income exceeded two-thirds of total household income, approximately three times that of part-time farms.

4.3 Results and Discussion

4.3.1 1106 Farms

4.3.1.1 Variable Description Statistics

Table 4.2 presents the definitions and means of variables. Overall, the percentage of serious work attitudes was not high. Only 23% of operators (including service providers and farmers) exhibited serious work attitudes during the harvest process. 59% of farmers purchased harvest outsourcing services. Less than half of the farmers used combine harvesters. In general, harvest outsourcing services and combine harvesting of rice are at moderate levels, and there is still room for further development. The average rice planting area was 0.33 ha, smaller than the average farm size in China (0.5 ha). In terms of individual characteristics of household heads, male household heads were dominant, which was consistent with the current reality in rural China. The average age of household heads was 54.12 years old, with an average of 7.01 years of schooling, equivalent to the junior high school education level.

Table 4.2 Summary and definition of variables (1106 farms)

Variable	Definition	Mean	Std. Dev.
Dependent variable			
WA	Dummy = 1 if operator's harvest attitude was serious, 0 otherwise	0.23	0.42
Core independent variable			
Ser	Dummy = 1 if farmer bought outsourcing service, 0 otherwise	0.59	0.49
Production and harvesting variable			
Com	Dummy = 1 if combine harvesting, 0 if segmented harvesting	0.46	0.50
Wea	Dummy = 1 if bad weather when harvesting, 0 if normal weather	0.16	0.37
Pest	No pest = 1, slight pests = 2, general or serious pests = 3	1.84	0.76
Area	Rice planting area (ha)	0.33	0.35
Flat	Dummy = 1 if plot terrain is flat, 0 otherwise	0.75	0.43
Htor	Distance from homestead to the nearest paved road	0.34	0.76
Labor	Dummy = 1 if farmer reported a lack of labor, 0 otherwise	0.28	0.45
Sav	Dummy = 1 if farmer picked up rice left in field, 0 otherwise	0.16	0.37
Price	Sale price of rice (CNY/kg)	2.98	0.34
Household and individual variable			
Gen	Gender of household head (male = 1, female = 0)	0.84	0.36
Age	Age of household head	54.12	10.63
Edu	School years of household head (years)	7.01	2.66
Train	Dummy = 1 if household head had agricultural training, 0 otherwise	0.09	0.29
Tinc	Household income (ten thousand CNY)	7.07	5.45
Rincs	Rice income as a percentage of total income (%)	15.80	18.16
N	Number of samples	1,106	

Notes The highest correlation coefficient between independent variables is no more than 0.8 and the VIF-value is less than 10, which means there is no multicollinearity
Data source Author's calculation based on the survey

Table 4.3 gives the average work attitudes of 1106 farms. As expected, farmers generally exhibited a more serious work attitude compared to service providers. Specifically, 32% of service providers demonstrated serious work attitudes, while 17% of farmers had serious work attitudes, nearly half of the former. The use of combine harvesters intensified the diligent work attitudes of both farmers and service providers, especially the service providers.

Table 4.3 Average work attitude of operators (1106 farms)

Sample	Work attitude (WA)
Average	0.23
Self-service (OS = 0)	0.32
Outsourcing service (OS = 1)	0.17
Outsourcing service using combine harvesting (OS = 1 and Com = 1)	0.21
Self-service using combine harvesting (OS = 0 and Com = 1)	0.33

Data source Author's calculation based on the survey

4.3.1.2 Estimation Results

Table 4.4 gives the estimation results of 1106 sample. Column (1) presents the marginal effects of the Logit regressions in Eq. (4.6) (without the cross term). The marginal effect of harvest outsourcing services (OS) is negative and significant (−0.342), which indicates that outsourcing services reduce the probability of a serious attitude. In other words, service providers are less serious about harvesting than farmers. This is consistent with our assumption that there exists a moral hazard, which diminishes the effort levels of service providers.

There are some factors that have negative effects on the probability of serious work attitudes. The marginal effects of bad weather (Wea) and pests (Pest) are significantly negative, which means that bad weather and pests reduce the probability of serious work attitudes. Both bad weather and pests make harvesting more difficult and increase harvest losses. In such cases, operators may increase the forward speed of machines to finish harvesting as soon as possible to avoid greater harvest losses caused by bad weather or pests. As a result, operators cannot be as cautious as usual. The marginal effects of household income (Tinc) and rice income share (Rincs) are negative and significant, implying that increased household income and rice income share will reduce the probability of serious work attitudes. However, these effects are very small.

Moreover, there are some factors that have positive effects on serious work attitudes. The marginal effect of combine harvesting (Com) is positive and significant. It means that combine harvesters increase the probability of serious work attitudes. The marginal effect of area (Area) is significantly positive, implying that increased planting area will increase the probability of serious work. For service providers, a larger planting area means higher service fees, which incentivizes them to maintain a serious work attitude. For farmers, expanding their planting area leads to more careful harvesting practices due to the increased importance of rice cultivation. The significantly positive marginal effect of land terrain (Flat) implies that operators are more likely to work effectively on flat farmland, which facilitates harvesting operations, especially mechanized ones. The significantly positive marginal effect of distance from homestead to the nearest paved road (Htor) means that the farther the distance from homestead to the nearest paved road, the more serious the operator's attitude

Table 4.4 Estimation results on work attitude (1106 farms)

	Operators' work attitude			
	(1) Without OS × Com		(2) With OS × Com	
Core independent variable				
OS	−0.342***	(0.04)	−0.407***	(0.06)
OS × Com			0.195*	(0.10)
Production and harvesting variables				
Com	0.144***	(0.04)	0.011	(0.08)
Wea	−0.094**	(0.04)	−0.093**	(0.04)
Pest = 2	−0.108***	(0.03)	−0.106***	(0.03)
Pest = 3	−0.115***	(0.03)	−0.110***	(0.03)
Area	0.221***	(0.05)	0.221***	(0.05)
Flat	0.105***	(0.03)	0.103***	(0.03)
Htor	0.048***	(0.01)	0.046***	(0.01)
Labor	−0.023	(0.03)	−0.025	(0.03)
Sav	0.057*	(0.03)	0.065**	(0.03)
Price	0.030	(0.04)	0.023	(0.04)
Household and individual variables				
Gen	0.014	(0.03)	0.009	(0.03)
Age	0.002*	(0.00)	0.002*	(0.00)
Edu	0.006	(0.00)	0.007	(0.00)
Train	−0.025	(0.05)	−0.027	(0.05)
Tinc	−0.005*	(0.00)	−0.005*	(0.00)
Rincs	−0.002**	(0.00)	−0.002**	(0.00)
Advantageous region	Yes		Yes	
Pseudo R^2	0.147		0.150	
N	1106		1106	
Lincom test			Odds ratio	
The effect of harvest outsourcing service among combine harvesting				
			0.241**	(0.13)
The effect of combine harvesting among harvest outsourcing service				
			4.001***	(1.67)

Notes (1) All coefficients are marginal effects estimates except the "Lincom" test. (2) ***, **, and * indicate 1, 5, and 10% significance levels, respectively. (3) Robust standard errors statistics are in parentheses. (4) Odds ratio is the ratio of odds. It is the ratio of the odds of WA = 1 for one group divided by the odds of WA = 1 for the other group. Odds ratio > 1 indicates increased occurrence of WA = 1, Odds ratio < 1 indicates decreased occurrence of WA = 1

Data source Author's calculation based on the survey

toward harvesting. Firstly, longer distance implies longer unpaved roads that require careful driving. Secondly, farmers living far from paved roads are typically older and economically disadvantaged (Dercon et al. 2009), placing a high value on food and a dislike for losses (Greeley 1986). They tend to take harvesting more seriously, as indicated by the positive marginal effects of food saving awareness (Sav) and age (Age). Farmers who are willing to pick up lost rice and older farmers typically have a stronger awareness of conserving food. They are reluctant to incur harvest losses, thereby increasing the probability of earnest harvesting.

Column (2) in Table 4.4 presents the marginal effects of the Logit regression in Eqs. (4.4–4.7) (with the cross term). In both segmented harvesting and combine harvesting, service providers are less serious than farmers. The marginal effect of harvest outsourcing services (OS) is negative and significant (−0.407). It means that the work attitudes of service providers using segmented harvesting are less serious than that of farmers using segmented harvesting. The marginal effect of cross term (OS $\times$ Com) is statistically significant, which implies that the effects of outsourcing services depend on the harvesting methods used. To compare the work attitudes of service providers using combine harvesting and farmers using combine harvesting, a "Lincom" test is performed in Stata. As shown at the bottom of Table 4.4, the odds ratio for harvest outsourcing services using combine harvesting relative to self-service using combine harvesting is less than 1 and statistically significant. This implies that service providers using combine harvesting have the less odds of serious attitude than farmers using combine harvesting. These results are in line with our previous analysis; that is, there is moral hazard in harvest outsourcing services, and service operators exhibit less serious work attitudes than farmers. The odds ratio for combine harvesting using outsourcing services relative to segmented harvesting using outsourcing services is more than 1 and statistically significant. This indicates that using combine harvesting increases the likelihood of the service operators having serious work attitudes. As mentioned earlier, combine harvesting can complete more harvesting stages than segmented harvesting, making service fees of combine harvesting higher than that of segmented harvesting (Poungchompu and Chantanop 2016). Therefore, with the incentive of higher service fees, service providers using combine harvesters are more likely to exhibit serious work attitudes compared to those using segmented harvesting.

4.3.2 *Farming Scale Perspective*

4.3.2.1 Variable Description Statistics

Table 4.5 presents definitions and means of variables for small-scale and large-scale farms. There were obvious differences between small-scale and large-scale farms in the use of harvest outsourcing services and combine harvesters. Large-scale farms were more likely to use harvest outsourcing services and combine harvesters. 73% of large-scale farms purchased harvest outsourcing services, which was 1.55 times

Table 4.5 Summary and definition of variables (farming scale)

Variable	Definition	Small-scale farms	Large-scale farms
Dependent variable			
WA	Dummy = 1 if operator's harvest attitude was serious, 0 otherwise	0.24	0.22
Core independent variable			
Ser	Dummy = 1 if farmer bought outsourcing service, 0 otherwise	0.47	0.73
Production and harvesting variable			
Com	Dummy = 1 if combine harvesting, 0 if segmented harvesting	0.38	0.57
Wea	Dummy = 1 if bad weather when harvesting, 0 if normal weather	0.12	0.20
Pest	No pest = 1, slight pests = 2, general or serious pests = 3	1.79	1.91
Area	Rice planting area (ha)	0.12	0.55
Flat	Dummy = 1 if plot terrain is flat, 0 otherwise	0.77	0.76
Htor	Distance from homestead to the nearest paved road	0.42	0.25
Labor	Dummy = 1 if farmer reported a lack of labor, 0 otherwise	0.31	0.23
Sav	Dummy = 1 if farmer picked up rice left in field, 0 otherwise	0.15	0.17
Price	Sale price of rice (CNY/kg)	3.04	2.92
Household and individual variable			
Gen	Gender of household head (male = 1, female = 0)	0.84	0.85
Age	Age of household head	55.43	52.80
Edu	School years of household head (years)	6.87	7.19
Train	Dummy = 1 if household head had agricultural training, 0 otherwise	0.11	0.08
Tinc	Household income (ten thousand CNY)	6.31	7.95
Rincs	Rice income as a percentage of total income (%)	7.82	24.32
N	Number of samples	548	532

Notes (1) The highest correlation coefficient between independent variables is no more than 0.8 and the VIF-value is less than 10, which means there is no multicollinearity
Data source Author's calculation based on the survey

that of small-scale farms (47%). The adoption rate of combine harvesters was not high. 57% of large-scale farms used combine harvesters, which was 1.50 times that of small-scale farms (38%).

The proportion of rice income to total income averaged 23.32% for large-scale farms, which was 3.11 times that of small-scale farms (7.82%). Therefore, large-scale farms were more willing to invest in the use of machinery than small-scale farms. Large-scale farms also had more capability to invest in the use of machinery than small-scale farms because their average total income was 16.4 thousand CNY higher than that of small-scale farms.

Moreover, large planting area of large-scale farms leads to a higher demand for combine harvesters. Service providers are more likely to provider service to large-scale farms. The average rice planting area varied greatly between small-scale farms and large-scale farms. The average rice planting area for small-scale farms was only 0.12 ha, while that for large-scale farms was 0.55 ha. It is well-known that the use of outsourcing services and combine harvesters can alleviate labor shortages. Thus, despite the fact that the planting area of large-scale farms was much larger than that of small-scale farms, large-scale farms experienced less labor shortage than small-scale farms, which may be the result of their high adoption of combine harvesters and outsourcing services.

The average age of household heads in large-scale farms was 52.80 years old, which was 2.63 years younger than household heads in small-scale farms. There was little difference in the schooling years between the two groups of household heads, with slightly higher educational years for household heads in large-scale farms compared those in small-scale farms.

Table 4.6 gives the average work attitudes among farms of different scales. On both small-scale farms and large-scale farms, farmers with serious attitudes toward work outnumbered service providers. On small-scale farms, only 12% of service providers had serious attitudes, while 34% of farmers exhibited serious work attitudes, nearly three time that of service providers. On large-scale farms, the average work attitude of service providers (0.20) was still lower than that of farmers (0.27), with smaller difference. Even when using combine harvesters, service providers were still less serious than farmers.

The average work attitude of farmers on small-scale farms (0.34) was more serious than that of farmers on large-scale farms (0.27). As the scale increases, it becomes more difficult for large-scale farmers to manage farming, leading to a

Table 4.6 Average work attitudes in different farm scales

Sample	Small-scale farms	Large-scale farms
Average	0.24	0.22
Self-service (OS = 0)	0.34	0.27
Outsourcing service (OS = 1)	0.12	0.20
Outsourcing service using combine harvesting (OS = 1 and Com = 1)	0.14	0.26
Self-service using combine harvesting (OS = 0 and Com = 1)	0.17	0.42

Data source Author's calculation based on the survey

decline in their effort levels. In contrast, service providers were more diligent when providing services to large-scale farms (0.20), compared to small-scale farms (0.12). The purpose of service providers is profitability. They will earn higher service fees by serving large-scale farms because the service fee is proportional to the serviced area. Therefore, service providers became more diligent when serving large-scale farms. This is also observed when service providers use combine harvesters.

Using combine harvesters made the service providers more serious about their work, especially on large-scale farms. This may be caused by the higher service fees for combine harvesting (Poungchompu and Chantanop 2016). When service providers provided combine harvesting services to large-scale farms, their average work attitude (0.26) was very close to that of farmers (0.27). Using combine harvesters also made large-scale farmers more serious. However, small-scale farmers using combine harvesters decreased their serious work attitudes (from 0.34 to 0.17), although it was still more serious than service providers. This may be due to the fact that combine harvesters are not suitable for small-scale farmland. Moreover, small-scale farmers own combine harvesters primarily to provide outsourcing services to other farmers. The benefit from reducing harvest losses in their own fields is much less than that from providing harvesting services to other farmers. Therefore, using combine harvesters reduced the diligent work attitude of small-scale farmers.

4.3.2.2 Estimation Results

Table 4.7 presents the marginal effects of factors for small-scale farms and large-scale farms (without the cross term). The marginal effects of outsourcing services (OS) for both small-scale farms (−0.283) and large-scale farms (−0.338) are negative and significant at the 1% significance level, implying that outsourcing services reduce the probability of a serious attitude. Whether providing services to small-scale or large-scale farms, service providers show a less serious work attitude compared to farmers.

The marginal effect of combine harvesting (Com) is positive and significant for large-scale farms, but not significant for small-scale farms. Compared to the tools used for segmented harvesting, combine harvester is large and suitable for large-scale farmland. Operating combine harvesters on small plots is difficult (Otsuka et al. 2016). In addition, the marginal effect of bad weather (Wea) becomes insignificant for large-scale farms. This may be due to the fact that large-scale farmers can use weather information to make scientific decisions about harvesting time, thereby mitigating the effect of weather. Meanwhile, due to higher service fees, service providers, especially those offering combine harvesting, are more willing to serve large-scale farms (Zhou 2017). As a result, large-scale farms can timely access mechanical harvesting services before bad weather arrives. This trend is evident in the higher adoption of outsourcing services and combine harvesting by large-scale farms (see Table 4.5). Furthermore, the marginal effect of sale price (Price) was negative and significant, meaning that an increase sale price of rice decreases the probability of diligent work attitudes among large-scale farms. On large-scale farms, most harvesting work is undertaken

Table 4.7 Estimation results on work attitude without cross term (farming scale)

	Operators' work attitude			
	(1) Small-scale farms		(2) Large-scale farms	
Core independent variable				
OS	−0.283***	(0.06)	−0.338***	(0.06)
Production and harvesting variables				
Com	0.011	(0.06)	0.179***	(0.05)
Wea	−0.419***	(0.13)	0.073	(0.06)
Pest = 2	−0.123***	(0.04)	−0.095**	(0.04)
Pest = 3	−0.102**	(0.05)	−0.150***	(0.04)
Area	0.044	(0.38)	0.125***	(0.05)
Flat	0.231***	(0.04)	−0.028	(0.04)
Htor	0.033*	(0.02)	0.104***	(0.03)
Labor	0.009	(0.03)	−0.066	(0.05)
Sav	0.095**	(0.05)	0.028	(0.05)
Price	0.057	(0.04)	−0.290***	(0.11)
Household and individual variables				
Gen	−0.018	(0.05)	0.054	(0.05)
Age	0.003*	(0.00)	0.002	(0.00)
Edu	0.010	(0.01)	0.001	(0.01)
Train	0.009	(0.06)	0.011	(0.07)
Tinc	−0.012	(0.01)	−0.003	(0.00)
Rincs	−0.007***	(0.00)	−0.001	(0.00)
Advantageous region	Yes		Yes	
pseudo R^2	0.286		0.157	
N	548		532	

Notes (1) All coefficients are marginal effects estimates. (2) ***, **, and * indicate 1, 5, and 10% significance levels, respectively. (3) Robust standard errors statistics are in parentheses
Data source Author's calculation based on the survey

by service providers. The increase in the sale price of rice leads to increased demand for harvest outsourcing services. Providing harvest outsourcing services to more farms inevitably reduces their probability of serious work attitudes.

Table 4.8 presents the marginal effects of the Logit estimations in Eq. (4.7) for different farm scales (with the cross term). For both small-scale and large-scale farms, the marginal effects of harvest outsourcing services (OS) are significantly negative at the 1% significance level. Consistent with the previous results, purchasing outsourcing services reduces the probability of service providers adopting a serious work attitude toward segmented harvesting.

At the bottom of Table 4.8, the "Lincom" test shows that the odds ratios of harvest outsourcing services using combine harvesting relative to self-service using combine

Table 4.8 Estimation results on work attitude with cross term (farming scale)

	Operator's work attitude			
	(1) Small-scale farms		(2) Large-scale farms	
Core independent variable				
OS	-0.328^{***}	(0.08)	-0.457^{***}	(0.09)
OS × Com	0.194	(0.18)	0.319^{**}	(0.13)
Production and harvesting variables				
Com	−0.137	(0.15)	−0.050	(0.10)
Wea	-0.421^{***}	(0.13)	0.090	(0.06)
Pest = 2	-0.122^{***}	(0.04)	-0.094^{**}	(0.04)
Pest = 3	-0.105^{**}	(0.05)	-0.141^{***}	(0.04)
Area	0.046	(0.38)	0.119^{**}	(0.05)
Flat	0.234^{***}	(0.04)	−0.039	(0.04)
Htor	0.032^{*}	(0.02)	0.105^{***}	(0.03)
Labor	0.007	(0.03)	−0.070	(0.04)
Sav	0.095^{**}	(0.05)	0.049	(0.05)
Price	0.052	(0.04)	-0.344^{***}	(0.11)
Household and individual variables				
Gen	−0.025	(0.05)	0.044	(0.05)
Age	0.003^{*}	(0.00)	0.002	(0.00)
Edu	0.010^{*}	(0.01)	0.002	(0.01)
Train	0.007	(0.06)	0.015	(0.07)
Tinc	−0.012	(0.01)	−0.003	(0.00)
Rincs	-0.007^{***}	(0.00)	−0.001	(0.00)
Advantageous region	Yes		Yes	
Pseudo R^2	0.288		0.169	
N	548		532	
Lincom test	Odds ratio		Odds ratio	
The effect of harvest outsourcing service among combine harvesting				
	0.340	(0.43)	0.377	(0.26)
The effect of combine harvesting among harvest outsourcing service				
	1.589	(1.11)	6.740^{***}	(3.56)

Notes (1) All coefficients are marginal effects estimates except the "Lincom" test. (2) ***, **, and * indicate 1, 5, and 10% significance levels, respectively. (3) Robust standard errors statistics are in parentheses. (4) Odds ratio is the ratio of odds. It is the ratio of the odds of WA = 1 for one group divided by the odds of WA = 1 for the other group. Odds ratio > 1 indicates increased occurrence of WA = 1, Odds ratio < 1 indicates decreased occurrence of WA = 1
Data source Author's calculation based on the survey

harvesting are less than 1 for both small-scale and large-scale farms, but insignificant. The higher service fee for combine harvesting makes service providers more diligent, thus mitigating moral hazard issue. For both small-scale and large-scale farms, the odds ratios for combine harvesting using outsourcing services relative to segmented harvesting using outsourcing services are more than 1, but it is only statistically significant for large-scale farms. For large-scale farms, the use of combine harvesters increases the probability of service providers maintaining a serious work attitude; however, it has no significant effect on small-scale farms. Operators of combine harvesters are more willing to provide services to large-scale farms (Zhou 2017; Qu et al. 2021b). Because the service fee, which is proportional to the serviced area, is higher. Meanwhile, combine harvesters are better suited for operations on large-scale farmland (Otsuka et al. 2016). Small planting area can pose operational difficulties for combine harvesters. The marginal effects of other factors are similar to those in Table 4.7.

4.3.3 Part-Time Farming Perspective

4.3.3.1 Variable Description Statistics

Table 4.9 gives the variables definitions and means for part-time farms and business farms. Part-time and business farms share many similarities with small-scale and large-scale farms. Overall, part-time and business farms do not differ as much as small-scale and large-scale farms above.

57% of business farms purchased harvest outsourcing services, whereas this proportion was 60% for part-time farms. Part-time farms allocate household labors between agriculture sector and non-agriculture sectors. Labor shortages in part-time farms may account for their high adoption of harvest outsourcing services. However, despite the higher adoption of outsourcing services among part-time farms, they still faced more severe labor shortages than business farms. 31% of part-time farms reported a lack of labor, while 23% of business farms indicated a labor shortage. 45% of part-time farms used combine harvesters, compared to a slightly higher proportion of 48% for business farms. Combine harvesters can harvest a large area of rice in a short period of time and are also more suitable for operating on large areas. The relatively higher adoption of combine harvesters by business farms is due to the fact that they had a larger rice planting area of 0.46 ha. In contrast, the average rice planting area for part-time farms was only 0.23 ha, which is half of that for business farms. Meanwhile, rice production was an important source of income for business farms. Rice income accounted for a quarter of the total income of business farms, which was 2.60 times that of part-time farms.

Table 4.10 gives that the average work attitudes of part-time farms and business farms. On both part-time farms and business farms, service providers' average work attitude was less serious than that of farmers. On part-time farms, 35% of farmers had serious work attitudes, more than twice that of service providers (15%). On business

Table 4.9 Summary and definition of variables (part-time farming)

Variable	Definition	Part-time farms	Business farms
Dependent variable			
WA	Dummy = 1 if operator's harvest attitude was serious, 0 otherwise	0.23	0.23
Core independent variable			
Ser	Dummy = 1 if farmer bought outsourcing service, 0 otherwise	0.60	0.57
Production and harvesting variable			
Com	Dummy = 1 if combine harvesting, 0 if segmented harvesting	0.45	0.48
Wea	Dummy = 1 if bad weather when harvesting, 0 if normal weather	0.14	0.18
Pest	No pest = 1, slight pests = 2, general or serious pests = 3	1.87	1.80
Area	Rice planting area (ha)	0.23	0.46
Flat	Dummy = 1 if plot terrain is flat, 0 otherwise	0.74	0.77
Htor	Distance from homestead to the nearest paved road (km)	0.28	0.41
Labor	Dummy = 1 if farmer reported a lack of labor, 0 otherwise	0.31	0.23
Sav	Dummy = 1 if farmer picked up rice left in field, 0 otherwise	0.14	0.19
Price	Sale price of rice (CNY/kg)	2.98	2.99
Household and individual variable			
Gen	Gender of household head (male = 1, female = 0)	0.85	0.84
Age	Age of household head	55.28	52.60
Edu	School years of household head (years)	7.06	6.93
Train	Dummy = 1 if household head had agricultural training, 0 otherwise	0.10	0.09
Tinc	Household income (ten thousand CNY)	7.03	7.13
Rincs	Rice income as a percentage of total income (%)	9.34	24.26
N	Number of samples	627	479

Notes (1) The highest correlation coefficient between independent variables is no more than 0.8 and the VIF-value is less than 10, which means there is no multicollinearity
Data source Author's calculation based on the survey

farms, the difference between farmers and service providers was smaller, but service providers with serious work attitudes (20%) were still fewer than farmers (28%). Even when using combine harvesters, service providers were still less serious than farmer.

Service providers were more serious when serving business farms compared to part-time farms. On part-time farms, 15% of service providers had serious attitudes, which was less than that on business farms (20%). The difference of planting area

Table 4.10 Average work attitudes in part-time farms and business farms

	Part-time farms	Business farms
Average	0.23	0.23
Self-service (OS = 0)	0.35	0.28
Outsourcing service (OS = 1)	0.15	0.20
Outsourcing service using combine harvesting (OS = 1 and Com = 1)	0.19	0.23
Self-service using combine harvesting (OS = 0 and Com = 1)	0.27	0.43

Data source Author's calculation based on the survey

between part-time farms and business farms may be the reason. As given in Table 4.9, business farms had larger rice planting areas than part-time farms. It means that service providers will earn higher income by servicing business farms. Motivated by the higher service fees, service providers are more likely to be serious when serving business farms.

However, part-time farmers (0.35) were more serious than business farmers (0.28). Despite higher farm income on business farms, it does not necessarily lead to a more serious work attitude among their owners.[5] This could be attributed to the challenges in managing agricultural production on larger planting areas of business farms, which often lead to extensive harvesting operations. Moreover, the average age of household heads on part-time farms was older than that of household heads on business farms (see Table 4.9), which might explain their more serious work attitudes, as elder farmers tend to understand the value of crops better.

The use of combine harvesters would enhance service providers' serious work attitudes. This could be due to the higher service fees associated with combine harvesting (Poungchompu and Chantanop 2016). Using combine harvesters encourages business farmers to be more serious, while part-time farmers are less so. This may stem from the suitability of combine harvesters for large fields on business farms versus smaller plots on part-time farms (Otsuka et al. 2016). Meanwhile, part-time farmers who own combine harvesters primarily provide outsourcing services to other farmers. The benefit saved from reducing harvest losses on their own fields is much less than the service fee earned from providing harvesting services to other farmers. Therefore, their serious work attitudes toward harvesting may decline accordingly.

4.3.3.2 Estimation Results

Table 4.11 gives the marginal effects of Eq. (4.6) (without the cross term). The marginal effect of outsourcing services (OS) is negative and significant for part-time farms (−0.368) and business farms (−0.345), which means that outsourcing services

[5] Nevertheless, this does not affect the fact that part-time farmers conduct extensive management. Because their human resource input is insufficient.

Table 4.11 Estimation results on work attitude without cross term (part-time farming)

	Operators' work attitude			
	(1) Part-time farms		(2) Business farms	
Core independent variable				
OS	-0.368^{***}	(0.05)	-0.345^{***}	(0.08)
Production and harvesting variables				
Com	0.122^{**}	(0.05)	0.188^{**}	(0.07)
Wea	-0.170^{***}	(0.06)	−0.051	(0.06)
Pest = 2	-0.122^{***}	(0.04)	-0.105^{**}	(0.04)
Pest = 3	-0.114^{**}	(0.05)	-0.109^{**}	(0.05)
Area	0.502^{***}	(0.12)	0.147^{***}	(0.05)
Flat	0.105^{**}	(0.04)	0.103^{**}	(0.05)
Htor	0.002	(0.02)	0.078^{***}	(0.02)
Labor	−0.038	(0.03)	−0.008	(0.04)
Sav	0.022	(0.05)	0.115^{**}	(0.05)
Price	0.047	(0.04)	0.042	(0.06)
Household and individual variables				
Gen	0.065	(0.05)	−0.053	(0.05)
Age	0.001	(0.00)	0.003	(0.00)
Edu	0.005	(0.01)	0.007	(0.01)
Train	−0.005	(0.06)	0.014	(0.07)
Tinc	−0.006	(0.01)	-0.007^{**}	(0.00)
Rincs	−0.003	(0.00)	−0.002	(0.00)
Advantageous region	Yes		Yes	
Pseudo R^2	0.154		0.204	
N	627		479	

Notes (1) All coefficients are marginal effects estimates. (2) ***, **, and * indicate 1, 5, and 10% significance levels, respectively. (3) Robust standard errors statistics are in parentheses
Data source Author's calculation based on the survey

reduce the probability of serious attitude and service providers are less serious than both part-time farmers and business farmers. This is consistent with our assumption that service providers have reduced effort levels, reflected in their less serious work attitudes.

Additionally, other factors also influence the work attitude of operators. The marginal effect of combine harvesting (Com) is positive and significant, suggesting that operators using combine harvesters are more serious compared to those using segmented harvesting. Bad weather (Wea) and pests (Pest) have significantly negative marginal effects, implying that bad weather and pests decrease the probability of operators' serious work attitudes, due to the challenges they pose to harvesting operations. The positive and significant marginal effects of rice planting area (Area) and terrain (Flat) mean that an increase in rice planting area and flat terrain increase the probability of operators' serious work attitudes. Large area and flat terrain favorable conditions for harvesting, especially for machinery operation. Therefore, operators exhibit a more serious work attitude when working on larger and flat farmland.

Table 4.12 gives the marginal effects of Eq. (4.7) (with the cross term). The marginal effect of outsourcing services (OS) is significantly negative for part-time farms (−0.478) and business farms (−0.350). It means that outsourcing services reduce the probability of serious work attitudes and service providers are less serious than household members when using segmented harvesting.

The results of "Lincom" tests are listed at the bottom of Table 4.12. For business farms, the odds ratio for harvest outsourcing services using combine harvesting relative to self-service using combine harvesting is 0.088, which is less than 1 and statistically significant. It means that service providers are less serious than business farmers when using combine harvesting. However, the odds ratio for part-time farms is less than 1, but not significant. On the one hand, this is because the combine harvesting increases the likelihood of serious work by service providers due to the income incentive. On the other hand, part-time farmers will conduct extensive management when using combine harvesters on their own farmland (as shown in Table 4.10). Part-time farmers own combine harvesters primarily to provide outsourcing services to other farmers. Compared to providing harvesting services to other farmers, the income gained from reducing harvest losses in their own fields is much lower. Therefore, part-time farmers may spend more time and effort providing outsourcing services to others than harvesting their own farmland. As a result, when using combine harvesters, there is no significant difference in the work attitudes of part-time farmers and service providers. The odds ratios for combine harvesting using outsourcing services relative to segmented harvesting using outsourcing services are more than 1 and statistically significant for both part-time farms (4.605) and business farms (3.985). It indicates that service providers using combine harvesters are more serious than those using segmented harvesting. This may be due to the fact that the service fee for combine harvesting is higher than that for segmented harvesting (Poungchompu and Chantanop 2016), which creates an incentive for service providers, thus increasing their likelihood of serious work attitudes.

4.4 Robustness Test

4.4.1 Propensity Score Matching

We study service provider's reduced effort level by comparing the work attitudes of farmers and service providers. However, farmers with more serious attitudes may choose to harvest on their own, which means that there may be potential self-selection bias when participating in harvest outsourcing services. Therefore, outsourcing services are regarded as the treatment group and self-services are regarded as the

Table 4.12 Estimation results on work attitude with cross term (part-time farming)

	Operators' work attitude			
	(1) Part-time farms		(2) Business farms	
Core independent variable				
OS	−0.478***	(0.08)	−0.350***	(0.11)
OS × Com	0.330**	(0.14)	0.012	(0.16)
Production and harvesting variables				
Com	−0.108	(0.10)	0.180	(0.12)
Wea	−0.172***	(0.06)	−0.051	(0.06)
Pest = 2	−0.121***	(0.04)	−0.105**	(0.04)
Pest = 3	−0.109**	(0.04)	−0.109**	(0.05)
Area	0.523***	(0.12)	0.147***	(0.05)
Flat	0.105**	(0.04)	0.103**	(0.05)
Htor	−0.000	(0.02)	0.078***	(0.02)
Labor	−0.045	(0.03)	−0.008	(0.04)
Sav	0.038	(0.05)	0.115**	(0.05)
Price	0.037	(0.04)	0.042	(0.06)
Household and individual variables				
Gen	0.046	(0.05)	−0.052	(0.05)
Age	0.001	(0.00)	0.003	(0.00)
Edu	0.006	(0.01)	0.007	(0.01)
Train	−0.008	(0.06)	0.014	(0.07)
Tinc	−0.006	(0.01)	−0.007**	(0.00)
Rincs	−0.002	(0.00)	−0.002	(0.00)
Advantageous region	Yes		Yes	
Pseudo R^2	0.164		0.204	
N	627		479	
Lincom test	Odds ratio		Odds ratio	
The effect of harvest outsourcing service among combine harvesting				
	0.362	(0.26)	0.088***	(0.08)
The effect of combine harvesting among harvest outsourcing service				
	4.605***	(2.63)	3.985*	(2.86)

Notes (1) All coefficients are marginal effects estimates except the "Lincom" test. (2) ***, **, and * indicate 1, 5, and 10% significance levels, respectively. (3) Robust standard errors statistics are in parentheses. (4) Odds ratio is the ratio of odds. It is the ratio of the odds of WA = 1 for one group divided by the odds of WA = 1 for the other group. Odds ratio > 1 indicates increased occurrence of WA = 1, Odds ratio < 1 indicates decreased occurrence of WA = 1

Data source Author's calculation based on the survey

comparison group. Propensity Score Matching (PSM) method is then used to test the robustness of the above results. Given the limited sample size and the small variability of the above results, PSM analysis is performed in the 1106 sample.

Matching greatly reduces differences between variables, as shown by the percentage reduction in bias in Table 4.13. An exception is the Train, which is already insignificant before matching. The bias after matching is less than 10% for each variable (Harder et al. 2010), and no variable is statistically different after matching.

After matching, there are 877 observations in the support region. Average treatment effect in Table 4.14 gives a decrease in both magnitude and significance level of the effect of outsourcing services on work attitudes, implying that untreated self-selection issues overestimate the effect of outsourcing services on work attitudes. However, service providers' average serious work attitude is still 0.186 lower than matched farmers. Therefore, PSM results also indicate the presence of moral hazard, which causes service providers to be less serious about harvesting work compared to farmers.

Table 4.13 Balancing test on matching variables

Variable	Mean		% Bias	% Reduction bias	*t*-test	
	Outsourcing service	Self-service			*t*	*p* value
Com	0.736	0.734	0.6	99.7	0.07	0.945
Htor	0.264	0.213	6.6	44.4	1.48	0.139
Area	0.329	0.343	−4.5	91.8	−0.82	0.413
Flat	0.795	0.810	−3.4	89.2	−0.61	0.542
Labor	0.234	0.256	−4.9	78.5	−0.85	0.397
Price	2.870	2.859	3.3	95.6	0.95	0.344
Gen	0.893	0.871	6.0	69.2	1.13	0.259
Edu	7.507	7.618	−4.2	85.9	−0.76	0.448
Train	0.103	0.089	5.1	−101.5	0.82	0.410
Rincs	15.718	14.394	7.8	85	1.50	0.135

Data source Author's calculation based on the survey

Table 4.14 Average treatment effect (ATT)

		Outsourcing service	Self-service	Difference
ATT	Matched	0.157	0.316	−0.186* (0.099)
Samples on support		542	335	

Note (1) ***, **, and * indicate 1, 5, and 10% significance levels, respectively. (2) Standard errors statistics are in parentheses. The standard error for the ATT is bootstrapped standard error of 500 replications. (3) We use nearest-neighbor matching ($n = 1$; caliper = 0.01) without replacement
Data source Author's calculation based on the survey

4.4.2 Regional Control Using Rice Cropping Regionalization

Another common regional classification in China divides rice production into six rice cropping regions (Mei et al. 1988; Xin and Li 2009). This division is based on the ecological environment (temperature, moisture, sunlight, altitude, and soil), socio-economic conditions (administrative division, population, land, and basic production conditions), and rice cultivation characteristics (rice cultivation system, variety type, tillage method, and cultivation techniques). The six cropping regions are as follows:

I. South China double rice cropping region: This region is dominated by hills and mountains. Rice is mainly distributed in coastal plains and inter-mountain basins, with a high multiple cropping index.
II. Central China double and single rice cropping region: Double and single rice cropping co-exist in this region. Double rice cropping is mainly distributed south area of the Yangtze River, while the single rice cropping is mainly found north area of the Yangtze River.
III. Southwestern plateau region of single and double rice cropping region: The topography in this region is complex. Rice is distributed in mountain basins, mountain dams, terraced fields, and barren ridges. Single rice cropping predominates in this region, which is characterized by a wide variety of pests and diseases.
IV. North China single rice cropping region: The region is dominated by single rice cropping. Natural disasters are frequent, and precipitation is unevenly distributed between years and seasons.
V. Northeast China early maturing and single rice cropping region: This region mainly grows single rice cropping. The terrain of this area is flat and open, with deep and fertile soil, which is suitable for the mechanization.
VI. Northwest China early maturing and single rice cropping region: Single rice cropping is grown in this region. The soil in this region is relatively infertile. Rice is mainly distributed in the basins and plains.

As given in Table 4.15, the corresponding surveyed provinces fall into five rice cropping regions based on the survey scope and geographical location.

Therefore, this rice cropping regionalization can also represent the characteristics of rice planting and harvest outsourcing services in China. The rice cropping regionalization is used to replace the control variable of the three rice advantageous regions above and conduct the robustness regression analysis.

Table 4.16 gives the estimation results of outsourcing services on work attitude with rice cropping reginal controls. The estimation results of the variables are very similar to those using rice advantageous regional controls in Table 4.4. Specifically, in column (1), the marginal effect of outsourcing services (OS) without cross term is -0.313, which is significant at the 1% significance level. It is close to the marginal effect of -0.342 for outsourcing services in Table 4.4, which is also significant at the 1% significance level. In column (2), the marginal effect of outsourcing services (OS) with cross term is -0.382 and significant at the 1% significance level. It is also

Table 4.15 Sample distribution based on rice cropping regionalization

Rice cropping regions	Sample
I. South China double rice cropping region	Guangxi, Guangdong, Fujian
II. Central China double and single rice cropping region	Yunnan, Guizhou, Hunan
III. Southwestern plateau region of single and double rice cropping region	Sichuan, Chongqing, Shannxi, Hubei, Jiangxi, Zhejiang, Jiangsu
IV. North China single rice cropping region	Anhui, Shandong, Tianjin
V. Northeast China early maturing and single rice cropping region	Liaoning, Jilin, Heilongjiang
VI. Northwest China early maturing and single rice cropping region	None

Data source Mei et al. (1988), Xin and Li (2009), and author's calculation based on the survey

Table 4.16 Estimation results on work attitude (rice cropping regional control)

	Operators' work attitude			
	(1) Without OS × Com		(2) With OS × Com	
Core independent variable				
OS	−0.313***	(0.04)	−0.382***	(0.06)
OS × Com			0.218*	(0.11)
Production and harvesting variables				
Com	0.119***	(0.04)	−0.034	(0.09)
Wea	−0.107**	(0.04)	−0.105**	(0.04)
Pest = 2	−0.117***	(0.03)	−0.116***	(0.03)
Pest = 3	−0.115***	(0.03)	−0.110***	(0.03)
Area	0.184***	(0.05)	0.182***	(0.05)
Flat	0.079**	(0.03)	0.077**	(0.03)
Htor	0.048***	(0.01)	0.047***	(0.01)
Labor	−0.021	(0.02)	−0.024	(0.02)
Sav	0.048	(0.03)	0.057*	(0.03)
Price	−0.021	(0.04)	−0.030	(0.04)
Household and individual variables				
Gen	−0.002	(0.03)	−0.007	(0.03)
Age	0.002	(0.00)	0.002	(0.00)
Edu	0.006	(0.00)	0.006	(0.00)
Train	−0.037	(0.05)	−0.040	(0.05)
Tinc	−0.006**	(0.00)	−0.005**	(0.00)

(continued)

Table 4.16 (continued)

	Operators' work attitude			
	(1) Without OS × Com		(2) With OS × Com	
Rincs	−0.002*	(0.00)	−0.002	(0.00)
Rice cropping region	Yes		Yes	
pseudo R^2	0.169		0.173	
N	1106		1106	
Lincom test			Odds ratio	
The effect of outsourcing services among combine harvesting				
			0.320*	(0.21)
The effect of combine harvesting among outsourcing services				
			3.625***	(1.53)

Notes (1) All coefficients are marginal effects estimates except the "Lincom" test. (2) ***, **, and * indicate 1, 5, and 10% significance levels, respectively. (3) Robust standard errors statistics are in parentheses. (4) Odds ratio is the ratio of odds. It is the ratio of the odds of WA = 1 for one group divided by the odds of WA = 1 for the other group. Odds ratio > 1 indicates increased occurrence of WA = 1, Odds ratio < 1 indicates decreased occurrence of WA = 1
Data source Author's calculation based on the survey

close to the marginal effect of −0.407 for outsourcing services in Table 4.4, which is significant at the 1% significance level. Moreover, the results using rice cropping regional controls are also similar to those using rice advantageous regional controls from farming scale and part-time farming perspective.

4.5 Summary

This chapter examines the moral hazard in harvest outsourcing services. The moral hazard discussed in this chapter refers to the reduction in effort levels by service providers when delivering harvesting services, where this effort level is lower than that of the farmers. The work attitude during harvesting is used to capture the effort levels of both farmers and service providers. Service providers are regarded to have reduced effort level if their work attitude is less serious than that of farmers; conversely, if their work attitude is not less serious than that of farmers, then service providers have no reduced effort level. Therefore, this chapter examines service providers' reduced effort level by comparing the work attitudes of farmers and service providers.

Compared to farmers, more service providers have a diligent work attitude. Logit regression results show that when using segmented harvesting and combine harvesting, the diligence of service providers is lower than that of farmers, thereby indicating a reduction in the effort level of service providers. Moreover, bad weather, pests, and increased household income and rice income share will decrease the probability of serious work attitudes, while increased planting area, flat farmland terrain, longer distance from homestead to the nearest paved road, farmer's saving attitude toward loss, and increased age of household head will increase the probability of serious work attitudes.

The effect of harvest outsourcing services on operators' work attitudes is also studied from the perspective of faming scale and part-time farming. The average work attitude of farmers was more serious than that of service providers. As the scale increases, the average work attitude of service providers became more serious, while that of farmers became less serious. Using combine harvesters made service providers and large-scale farmers have a more serious work attitude. However, the average work attitude of small-scale farmers became less serious when using combine harvesters. Logit regressions show that for both small-scale farms and large-scale farms, when using segmented harvesting, service providers' work attitude is less serious compared to farmers. However, such a difference in work attitudes of service providers and farmers is not observed in combine harvesting. Moreover, on large-scale farms, using combine harvesters increases the probability of serious work attitudes by service providers.

The results from part-time farms and business farms are similar to those from small-scale farms and large-scale farms. For both part-time farms and business farms, service providers were less serious than farmers. When serving business farms, service providers were more serious than serving part-time farms. However, business farmers exhibited less seriousness than part-time farmers when harvesting their own farmland. When using combine harvesters, the average work attitudes of service providers and business farmers became more serious, while that of part-time farmers became less serious. Logit regressions show that, for business farms, service providers are less serious than farmers both in segmented harvesting and combine harvesting. On part-time farms, this difference is only observed when using segmented harvesting. Moreover, combine harvesting increases the probability of serious work attitudes by service providers on both part-time farms and business farms.

To mitigate possible self-selection biases, we regard outsourcing service as the treatment group and self-service as the comparison group and use PSM to test the robustness of the above results. The results of PSM also support the finding that service providers' work attitude is less serious than that of farmers. Furthermore, robustness tests using rice cropping regional controls yield results consistent with those obtained using rice advantageous regional controls.

In this chapter, service providers' moral hazards have been demonstrated by their less serious work attitudes than farmers. It contributes to the literature by providing the first empirical examination of moral hazards in harvest outsourcing services.

References

Alizadeh MR, Allameh A (2013) Evaluating rice losses in various harvesting practices. Int Res J Apllied Basic Sci 4:894–901

Azam JP, Gubert F (2006) Migrants' remittances and the household in Africa: a review of evidence. J Afr Econ 15:426–462. https://doi.org/10.1093/jae/ejl030

Belton B, Fang P, Reardon T (2018) Mechanization outsourcing services in Myanmar's dry zone. Michigan State University, Michigan, USA

Breustedt G, Glauben T (2007) Driving forces behind exiting from farming in Western Europe. J Agric Econ 58:115–127. https://doi.org/10.1111/j.1477-9552.2007.00082.x

Cai J, Liu WY (2019) Agricultural social service and opportunistic behavior: Take agricultural machinery operation services as example. Reform, 18–29

Chen C, Tang R Di (2020) The impact of rice production outsourcing on farmland renting: Based on the analysis of farming scale heterogeneities. J Nanjing Agric Univ Soc Sci Ed 20:5–24. https://doi.org/10.19714/j.cnki.1671-7465.2020.0085

Coelli T, Battese G (1996) Identification of factors which influence the technical inefficiency of Indian farmers. Aust J Agric Econ 40:103–128. https://doi.org/10.1111/j.1467-8489.1996.tb00558.x

De Brauw A, Rozelle S (2008) Migration and household investment in rural China. China Econ Rev 19:320–335. https://doi.org/10.1016/j.chieco.2006.10.004

De Xu, Deng X, Guo SL, Liu SQ (2019) Labor migration and farmland abandonment in rural China: empirical results and policy implications. J Environ Manage 232:738–750. https://doi.org/10.1016/j.jenvman.2018.11.136

Deng X, De XuD, Zeng M, Bin QY (2020) Does outsourcing affect agricultural productivity of farmer households? evidence from China. China Agric Econ Rev 12:673–688. https://doi.org/10.1108/CAER-12-2018-0236

Dercon S, Gilligan DO, Hoddinott J, Woldehanna T (2009) The impact of roads and agricultural extension on consumption growth and poverty in fifteen Ethiopian villages. Am J Agric Econ 91:1007–1021. https://doi.org/10.1111/j.1467-8276.2009.01325.x

Diener E, Suh EM, Lucas RE, Smith HL (1999) Subjective wellbeing: three decades of progress. Psychol Bull 125:276–302. https://doi.org/10.1037/0033-2909.125.2.276

Fan S, Hazell P, Thorat S (2000) Government spending, growth and poverty in rural India. Am J Agric Econ 82:1038–1051. https://doi.org/10.1111/0002-9092.00101

Greeley M (1986) Food, thchnology and employment: the farm-level post-harvest system in developing countries. J Agric Econ 37:333–347. https://doi.org/10.1111/j.1477-9552.1986.tb01602.x

Hao HG, Bin LX, Zhang JP (2013) Impacts of part-time farming on agricultural land use in ecologically-vulnerable areas in North China. J Resour Ecol 4:70–79. https://doi.org/10.5814/j.issn.1674-764x.2013.01.010

Harder VS, Stuart EA, Anthony JC (2010) Propensity score techniques and the assessment of measured covariate balance to test causal associations in psychological research. Psychol Methods 15:234–249. https://doi.org/10.1037/a0019623

Hu Y, Li B, Zhang Z, Wang J (2022) Farm size and agricultural technology progress: evidence from China. J Rural Stud 93:417–429. https://doi.org/10.1016/j.jrurstud.2019.01.009

Huan ML, Hou YX (2020) Quality control contract model of service in agricultural production outsourcing. J Agro-Forestry Econ Manag 19:288–296. https://doi.org/10.16195/j.cnki.cn36-1328/f.2020.03.31

Huang Y, Luo X (2020) How does cross-regional operation affect agricultural machiney service acquisition? J Huazhong Agric Univ (Social Sci Ed), 89–97

Ji XQ, Rozelle S, Huang JK et al (2016) Are China's farms growing? China World Econ 24:41–62. https://doi.org/10.1111/cwe.12143

Ji C, Guo HD, Jin SQ, Yang J (2017) Outsourcing agricultural production: evidence from rice farmers in zhejiang province. PLoS ONE 12:1–16. https://doi.org/10.1371/journal.pone.0170861

Kimhi A (2000) Is part-time farming really a step in the way out of agriculture? Am J Agric Econ 82:38–48. https://doi.org/10.1111/0002-9092.00004
Li Q, Huang JK, Luo RF, Liu CF (2013a) China's labor transition and the future of China's rural wages and employment. China World Econ 21:4–24. https://doi.org/10.1111/j.1749-124X.2013.12019.x
Li Q, Lin GH, He J (2013b) A correlation study on farmer's concurrent business behavior and changes in factors of production: Analysis based on a survey of farmers form rural fixed observation points. J Nanjing Agric Univ Soc Sci Ed 13:27–32
Li B, Qian Y, Kong F (2023) Does outsourcing service reduce the excessive use of chemical fertilizers in rural China? the moderating effects of farm size and plot size. Agriculture 13:1869. https://doi.org/10.3390/agriculture13101869
Li XF, Huang D, Wu LP (2019) Study on grain harvest losses of different scales of farms—Empirical analysis based on 3251 farmers in China. China Soft Sci, 184–192
Liu YH (2021) A study on the differences of agricultural production efficiency of farmers of different scales and influencing factors: an empirical analysis based on DEA-Tobit model. Ecol Econ 37:113–118
Lu QA, Du XD (2020) The outsourcing choice of agricultural production tasks: implications for food security – A multiple-task based approach. In: The 2020 Agricultural & Applied Economics Association Annual Meeting. Kansas City, Missouri
Lucas RE Jr (1967) Adjustment costs and the theory of supply. J Polit Econ 75:321–334. https://doi.org/10.1086/259289
MARA (2008) Regional layout planning of national superior agricultural products. http://www.moa.gov.cn/nybgb/2008/djiuq/201806/t20180611_6151652.htm. Accessed 17 Jan 2022
Mei F, Wu X, Yao C et al (1988) Rice cropping regionalization in China. Chinese J Rice Sci 2:97–110
Murphy S (2012) Changing perspectives: Small-scale farmers, markets and globalization. International Institute for Environment and Development and Hivos, London/The Hague
NBSC (2019) China agricultural machinery industry yearbook 2019. China Statistics Press, Beijing
NBSC (2021) 2020 Migrant worker monitoring survey report. http://www.stats.gov.cn/tjsj/zxfb/202104/t20210430_1816933.html. Accessed 15 Mar 2022
Otsuka K (2013) Food insecurity, income inequality, and the changing comparative advantage in world agriculture. Agric Econ (United Kingdom) 44:7–18. https://doi.org/10.1111/agec.12046
Otsuka K, Liu Y, Yamauchi F (2016) The future of small farms in Asia. Dev Policy Rev 34:441–461. https://doi.org/10.1111/dpr.12159
Pandey V, Shanoyan A, Peterson HC, Ross RB (2013) Principal-agent governance mechanism in an emerging biofuels supply chain in USA. Asian J Agric Rural Dev 3:532–542
Pfeffer MJ (1989) Part-time farming and the stability of family farms in the Federal Republic of Germany. Eur Rev Agric Econ 16:425–444. https://doi.org/10.1093/erae/16.4.425
Pingali P (2007) Agricultural mechanization: adoption patterns and economic impact. Handb Agric Econ 3:2779–2805. https://doi.org/10.1016/S1574-0072(06)03049-0
Poungchompu S, Chantanop S (2016) Economic aspects of rice combine harvesting service for farmer in Northeast Thailand. Asian Soc Sci 12:201–211. https://doi.org/10.5539/ass.v12n8p201
Qu X, Kojima D, Nishihara Y et al (2021a) Can harvest outsourcing services reduce field harvest losses of rice in China? J Integr Agric 20:1396–1406. https://doi.org/10.1016/s2095-3119(20)63263-4
Qu X, Kojima D, Nishihara Y et al (2021b) A study of rice harvest losses in China : Do mechanization and farming scale matter? Japanese J Agric Econ 23:83–88. https://doi.org/10.18480/jjae.23.0_83
Ruttan VW (2000) Technology, growth and development: an induced innovation perspective. Oxford University Press, New York
Sheng Y, Song LG, Yi Q (2017) Mechanisation outsourcing and agricultural productivity for small farms: Implications for rural land reform in China. In: Song L, Garnaut R, Cai F, Johnston L

(eds) China's new sources of economic growth: human capital, innovation and technological change, vol 2. Australian National University. Acton, Australia, pp 289–313
Smith A (1937) The wealth of nations: an inquiry into the nature and causes of the wealth of nations. University of Oregon, Eugene, USA
Weiss CR (1999) Farm growth and survival: econometric evidence for individual farms in Upper Austria. Am J Agric Econ 81:103–116. https://doi.org/10.2307/1244454
Wilk J, Andersson L, Warburton M (2013) Adaptation to climate change and other stressors among commercial and small-scale South African farmers. Reg Environ Chang 13:273–286. https://doi.org/10.1007/s10113-012-0323-4
Xin L, Li X (2009) Changes of multiple cropping in double cropping rice area of Southern China and its policy implications. J Nat Resour 24:58–65
Yang J, Huang ZH, Zhang XB, Reardon T (2013) The rapid rise of cross-regional agricultural mechanization services in China. Am J Agric Econ 95:1245–1251. https://doi.org/10.1093/ajae/aat027
Zhang ZM, Qian WR (2008) Analyses on farmers' behaviors of production in different scale of land management: based on the field survey in the middle and lower reaches of Yangtze River. J Sichuan Univ (Philosophy Soc Sci Ed), 87–93. https://doi.org/10.3969/j.issn.1006-0766.2008.01.012
Zhang XB, Yang J, Thomas R (2017) Mechanization outsourcing clusters and division of labor in Chinese agriculture. China Econ Rev 43:184–195. https://doi.org/10.1016/j.chieco.2017.01.012
Zhang QQ, Yan BB, Huo XX (2018) What are the effects of participation in production outsourcing? evidence from Chinese apple farmers. Sustain 10:4525. https://doi.org/10.3390/su10124525
Zhou J (2017) The dual division of agricultural socialization service and its consequences under the background of land circulation. J Nanjing Agric Univ (Social Sci Ed) 17:141–151

Chapter 5
The Effect of Outsourcing Service on Rice Harvest Loss Through Moral Hazard

In Chap. 4, we compare the work attitudes of service providers and farmers to study the moral hazards of service providers and find that service providers' work attitude is less serious than that of farmers. In this chapter, we will study whether the negative effect of harvest outsourcing services on operators' work attitudes will lead to increased rice harvest losses.

In this chapter, we will examine the path of harvest outsourcing services on rice harvest losses through moral hazards. Mediation analysis models are applied to test that moral hazards—measured as whether the work attitudes of harvest operators are serious—mediate the effect of harvest outsourcing services on rice harvest losses.

5.1 Introduction

Machinery is an essential factor affecting harvest losses. Higher levels of mechanization can contribute to harvest losses due to technical factors or maintenance (Kantor et al. 1997; Parfitt et al. 2010). Both field experiments (Greeley 1982; Huang et al. 2018) and household surveys (Li et al. 2020) indicated that mechanized harvesters cause greater losses compared to manual practices. This aspect is even more significant in light of the irreversible trend toward agricultural mechanization in China.

It is noteworthy that most of mechanical harvesting works in China are undertaken by harvest outsourcing service organizations and individuals (Ji et al. 2017; Yang et al. 2013; Zhang et al. 2017). In 2020, there were a total of 194,600 agricultural machinery service organizations nationwide, including 78,900 agricultural machinery cooperatives; there were 40.08 million households engaged in providing agricultural machinery services, including 4.23 million agricultural machinery specialized service households.

X. Qu et al., *Harvest Loss in China*, The University of Tokyo Studies on Asia,
https://doi.org/10.1007/978-981-97-9156-9_5

Unlike outsourcing services in the secondary and tertiary industries, the concerning moral hazard in harvest outsourcing services has received little attention. However, careless mechanical harvesting by service providers could also be responsible for increased harvest losses (Chegere 2017). In China, farmers pay service fees (based on harvested area) to organizations or individuals who own machine and entrust them to complete the harvesting (Ji et al. 2017; Sun et al. 2018; Yang et al. 2013). Harvest outsourcing services not only present the substitution of machinery for labor, but also the substitution of market labor for household labor (Deng et al. 2020; Picazo-Tadeo and Reig-Martínez 2006; Sun et al. 2018; Zhang et al. 2017), in which farmers are principals and service providers are agents. There are inconsistent goals and information asymmetry between farmers and service providers, which will trigger moral hazard issues on agent's part (Cai and Liu 2019; Huan and Hou 2020). To maximize their utility function, service providers may increase forward speed of machines or leave plants in the corner unharvested. However, these actions will adversely affect farmers' utility function (Cai and Liu 2019; Pandey et al. 2013); the direct manifestation is increased harvest losses. Although farmers can observe harvest losses after harvesting, the retrospective observation fails to play an effective role in supervision. Harvest outsourcing service market in China remains a seller's market (Shen et al. 2015). Disputes over service quality frequently occur (Han 2019), often placing farmers at a disadvantage. Therefore, we contend that moral hazards in harvest outsourcing services may result in increased harvest losses.

Although a considerable body of studies have focused on the background of harvest outsourcing service (Yang et al. 2013; Zhang et al. 2017), farmers' willingness to outsource (Ji et al. 2017; Sun et al. 2018), and its effect on agriculture (Deng et al. 2020; Lu and Du 2020; Picazo-Tadeo and Reig-Martínez 2006), only a few studies have studied the principal–agent problem in harvest outsourcing services, and these studies have remained in a simple discussion or theoretical analysis (Cai and Liu 2019; Pandey et al. 2013). Different from the previous studies, this study is based on the principal–agent theory and explores the effect of outsourcing services on agriculture from the perspective of harvest losses. It utilizes mediation analysis to verify the influence of harvest outsourcing services on harvest losses through moral hazard.

This chapter focuses on how rice harvest losses are affected by harvest outsourcing services through moral hazard. Mediation analysis models are applied to test the hypothesis that moral hazards—measured by whether the work attitudes of harvest operators are serious—mediate the effect of harvest outsourcing services on rice harvest losses.

5.2 Data and Method

5.2.1 Data and Variable

The data used in this chapter also come from the two surveys presented in Chap. 3[1]—"National Rural Fixed Observation Point Survey" and "Investigation and Evaluation of Rice Harvest Loss". The sample consists of 1106 households.

5.2.1.1 Dependent Variable

Using the same method for measuring harvest losses as described in Chap. 3, harvest losses refer to losses that occur from the field to the storage places, including reaping loss, threshing loss, winnowing loss, and field transport loss. Questionnaires were used to elicit estimates of these four stages from farmers. Harvest losses are calculated as follow:

$$\text{HLR} = \frac{\text{harvest losses}}{\text{harvest losses} + \text{PRO}} = \frac{L_{\text{reap}} + L_{\text{thr}} + L_{\text{win}} + L_{\text{tra}}}{\left(L_{\text{reap}} + L_{\text{thr}} + L_{\text{win}} + L_{\text{tra}}\right) + \text{PRO}} \times 100\% \tag{5.1}$$

5.2.1.2 Core Independent Variable

The core independent variable is the harvest outsourcing service. As described in Chap. 4, outsourcing service is represented by the dummy variable OS. As shown in Eq. (5.2), if farmers undertook the harvesting work themselves, OS equals 0; if service providers undertook the harvesting work, OS equals 1.

$$\text{OS} = \begin{cases} 0, & \text{if farmers did not use harvest outsourcing services} \\ 1, & \text{if farmers used harvest outsourcing services} \end{cases} \tag{5.2}$$

5.2.1.3 Mediation Variable

As analyzed above, the effect of harvest outsourcing service on rice harvest losses is driven by service providers' moral hazards (reduced effort level). Therefore, we need to examine whether service providers have moral hazards (reduced effort level).

[1] Please refer to Chap. 3 for more information about the dataset.

Similar to the method in Chap. 4,[2] work attitude is used to measure operators' effort level. Operator's work attitude is estimated by farmers. The professional investigators from RFOP asked farmers to rate operators' work attitude, which is measured on a three-point scale of fine, general, and rough. The estimation of operator's work attitude by farmers is based on their observations during harvesting process, such as harvest losses, forward speed of machines, and other operations they observed. Although work attitude is a subjective variable, reliable estimation of subjective variable is an acceptable option in capturing abstract concepts in the absence of objective variables. We believe that farmers' estimates are reliable in large sample because they are the most knowledgeable about the harvesting. Similar subjective variables, such as happiness and satisfaction, are often used in sociological studies (Diener et al. 1999).

For simplicity, work attitude takes value of 1 if the answer was "fine", it means that operator's work attitude was serious. Work attitude takes value of 0, if the answer was "general" or "rough", it means that operator's work attitude was not serious. As shown in Eq. (5.3), dummy variable "WA" is used to denote work attitude.

$$\mathrm{WA} = \begin{cases} 0, & \text{if service provider/farmer's work attitude was not serious} \\ 1, & \text{if service provider/farmer's work attitude was serious} \end{cases} \tag{5.3}$$

5.2.1.4 Covariates[3]

The harvest loss rate has been found to be affected by various factors (Fenn and Laycock 2017; Qu et al. 2020). To avoid the possible omitted variables bias, three types of control variables are included based on our analysis and existing research: (1) production and harvesting conditions, (2) household and individual characteristics, and (3) regional control variables.

Production and Harvesting Conditions[4]

The first type of control variables is production and harvesting conditions, which includes harvest methods (Com), mechanical winnowing (Win), mechanical field transport (Tra), weather conditions (Wea), pest and disease conditions (Pest), rice planting area (Area), yield per hectare (Yield), land terrain (Flat), distance from the field to the storage places (Dis), labor shortage (Labor), farmers' attitude toward harvest losses (Sav), maturity state of rice (Mat), and sale price of rice (Price).

Harvest Methods (Com) Harvest losses may vary with harvest methods. Although combine harvester technology is more advanced, some field experiments (Huang et al.

---

[2] Please refer to Chap. 4 for more information.

[3] Here we explain the covariates affecting harvest loss rates. Covariates that affecting work attitudes are the same as those in Chap. 4. Please refer to Chap. 4 for more information.

[4] Please refer to Chap. 4 for the definitions and data collections of these variables. Here we only analyze the expected impact of these variables on harvest losses.

2018; Li et al. 1991) and household survey (Zhan 1995) indicated that harvest losses by combine harvesting were higher than those by segmented harvesting. Because combine harvesters are largely influenced by mechanical performance and operators' skills. Large combine harvesters are difficult to work effectively on small-scale farmland. However, opposite view exists (Gao et al. 2016). The measurement of harvesting method is the same as that in Chap. 4.

Mechanical Winnowing (Win) Rice winnowing could be done by sieving, wind, hand picking, and winnowing machine (Baloch 1999; Danbaba et al. 2019). Dummy variable Win is used to capture whether the machine was used in winnowing stages. If machine was used, Win equals 1, otherwise 0.

Mechanical Field Transport (Tra) The harvested rice is transported from field to storage places by humans, animals, or tractors (Bala et al. 2010). Transportation by machines can shorten the transportation time, thereby reducing field transport losses. Dummy variable Tra is used to capture whether the machine was used in field transport stages. If machine was used for field transport, Tra equals 1, otherwise 0.

Weather Conditions (Wea) Weather condition is the first factor that comes into mind when considering the cause of harvest losses. On the one hand, bad weather itself will increase losses during harvesting. On the other hand, in case of bad weather, harvesting may be accelerated or roughly completed, which will increase harvest losses. The measurement of weather is the same as that in Chap. 4.

Pest and Disease Conditions (Pest) Similar to bad weather, encountering pest and disease will increase harvest losses. The measurement of pest and disease condition is the same as that in Chap. 4.

Rice Planting Area (Area) Rice planting area is also a crucial factor affecting harvest losses (Cao et al. 2018). Planting area is related to the farming management as well as the operation of machine. Both are associated with harvest losses. The income benefits from large-scale farmland are great. Farmers are more willing to invest in production materials, such as pesticides and machinery usage, which can reduce harvest losses. In addition, large-scale farmland also facilitates mechanical harvesting.

Yield per Hectare (Yield) It has also been shown that harvest losses are related to the variety of the crop (Chuan-udom and Chinsuwan 2010), because different varieties have different tolerance to bad conditions (Smale et al. 2008; Widawsky and Rozelle 1998). Some varieties are highly resistant to lodging (Wang et al. 2021), while others have seeds that fall easily when ripe (Chuan-udom and Chinsuwan 2010). Yield is one of the most important traits in rice. Therefore, the yield of rice is used here to capture the varieties.

Land Terrain (Flat) Uneven terrain can affect harvest losses (Goldsmith et al. 2015). It largely determines the availability of machinery, which in turn affects whether farmers can outsource harvesting as well as the harvesting quality (Lu and Du 2020). The measurement of land terrain is the same as that in Chap. 4.

Distance from the Field to the Storage Places (Dis) Losses happen during the transport from field to storage places are included in harvest losses in this study. The distance from field to storage places is an important factor affecting the transportation. The longer the distance transported, the greater the possibility of losses.

Labor Shortage (Labor) Harvesting is labor-intensive. The lack of labor may lead to delayed harvesting and increased losses (Basavaraja et al. 2007; Hasan et al. 2020). Wu et al. (2017) found that lack of labor could amplify rice harvest losses, while sufficient labor was not effective in reducing losses. The measurement of labor shortage is the same as that in Chap. 4.

Farmers' Attitude Toward Harvest Losses (Sav) Farmers' attitudes toward harvest losses may affect harvest losses. Some farmers who value their food will be more careful when harvesting or stay at the edge of the field while service providers are working. These farmers may also invest more effort and money in managing pests and diseases in the early stages of rice cultivation. Therefore, farmers with saving attitudes toward rice may decrease harvest losses. The measurement of farmers' attitude toward harvest losses is the same as that in Chap. 4.

Maturity State of Rice (Mat) Harvest time is another important factor affecting harvest losses. If the rice is immature when harvesting, the high tensile strength of rice and stalks makes threshing difficult, thus increasing losses. If the rice is overripe when harvesting, the plant activity decreases and the connection strength between seeds and stalks becomes low. Rice is easily dropped by external forces, thus increasing losses when reaping (Wang et al. 2016). Therefore, losses increase regardless of whether the rice is immature or overripe. Dummy variable Mat takes the value of 1 if farmers reported the rice to be mature when harvesting, and 0 otherwise.

Sale Price of Rice (Price) The sale price of rice is used to measure the opportunity cost of farmers' losses. When the price is high, farmers will harvest more intensively (Minor et al. 2020; Qu et al. 2021). Conversely, when prices are too low, harvesting becomes uneconomical and farmers may abandon harvesting their crops, leading to increased losses (ReFED 2016). The measurement of rice sale price is the same as that in Chap. 4.

Moreover, the other two types of control variables ((2) household and individual characteristics; (3) regional control variables) are the same as those in Chap. 4.

5.2.2 Mediation Analysis Model

5.2.2.1 Theoretical Model

Mediation analyses typically investigate whether the association between two variables, X and Y, can reasonably be accounted for by a mediation variable, M (Hayes and Preacher 2010). Baron and Kenny's (1986) widely cited mediation analysis model explored such associations using three-variable multiple linear regression

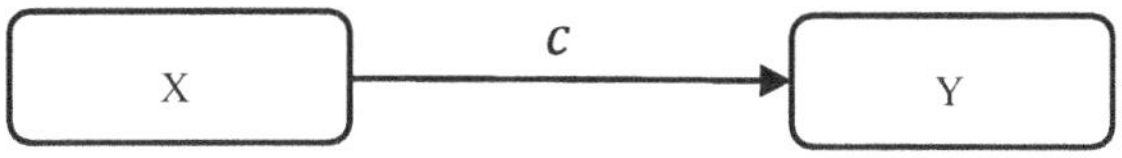

Fig. 5.1 The unmediated model

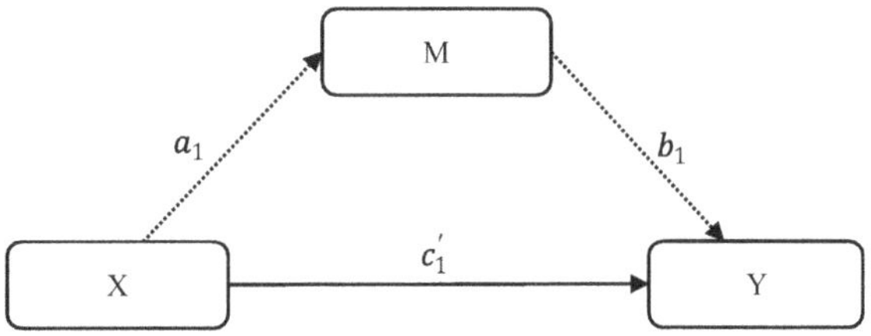

Fig. 5.2 The mediated model

models. In Figs. 5.1 and 5.2, there are four key paths, a_1, b_1, c_1, and c_1'. Figure 5.1 describes the unmediated association between X and Y as shown in Eq. (5.4). The solid line (path c_1) represents the *total effect* of X on Y. Figure 5.2 similarly summarizes the two equations required to model the indirect association mediated by M—Eqs. (5.5) and (5.6). The solid line (path c') indicates the *direct effect* of X on Y, adjusted for the effects of the mediator in Eq. (5.5). The dotted lines (path a_1 and path b_1) illustrate the mediation paths from X to Y, which is driven by M.

$$Y = c_0 + c_1 X + e_Y \tag{5.4}$$

$$Y = b_0 + b_1 M + c_1' X + e_{Y'} \tag{5.5}$$

$$M = a_0 + a_1 X + e_M \tag{5.6}$$

Y is the ultimate outcome, X is the focal predictor, and M is the mediator. c_i, b_i, and c_i represent regression estimates. e_Y, $e_{Y'}$, and e_M represent the error terms, which meet the standard assumptions of regression (i.e., homoscedasticity, normality, and independence).

The *indirect effect* (also called *mediation effect*) is the amount that Y is expected to change as X changes by one unit as a result of X's effect on M which, in turn, affects Y. It is quantified as the difference in coefficients method (i.e., $c_1 - c_1'$) and the product of coefficients method (i.e., $a_1 \times b_1$). For linear regression models, these two methods produce identical results.

5.2.2.2 Empirical Model

The harvested output that we usually discuss constitutes a (large) part of the total production. Another (small) part of production fails to be harvested due to environmental or technological factors; it is left in the field, resulting in harvest losses (Parfitt et al. 2010; Segrè et al. 2014). Therefore, as the adjustment of inputs can affect yield, a change in harvest conditions will affect the magnitude of harvest losses (Aulakh and Regmi 2013; Qu et al. 2021). Multiple regression has already been applied to examine factors influencing harvest losses (Basavaraja et al. 2007; Begum et al. 2012; Martins et al. 2014).

Based on Eq. (5.1), the dependent variable is theoretically restricted to 0–100%. In the sample, some farmers estimated their rice harvest loss rates to be 0%, which means that the data is left censoring. Therefore, Tobit model, also called a censored regression model (Tobin 1985) is used for estimation. Tobit model supposes that there is a latent variable, which linearly depends on independent variables. To investigate the effect of harvest outsourcing services on rice harvest loss rates with and without mediator, we consider a farm household model with the following relationship:

$$\mathrm{HLR}^* = c_0 + c_1(\mathrm{OS}) + \sum_{i=2}^{k} c_i z_i + e_1$$
$$\mathrm{HLR} = \begin{cases} \mathrm{HLR}^*, & \text{if } 0\% \leq \mathrm{HLR}^* \leq 100\% \\ 0, & \text{otherwise} \end{cases} \tag{5.7}$$

$$\mathrm{HLR}^* = b_0 + b_1(\mathrm{WA}) + c_1'(\mathrm{OS}) + \sum_{i=2}^{k} b_i z_i + e_2$$
$$\mathrm{HLR} = \begin{cases} \mathrm{HLR}^*, & \text{if } 0\% \leq \mathrm{HLR}^* \leq 100\% \\ 0, & \text{otherwise} \end{cases} \tag{5.8}$$

where HLR which is observed value equals HLR^* when $0\% \leq \mathrm{HLR}^* \leq 100\%$. HLR^* is latent variable which satisfied the classical linear model assumption. c_1 captures the total effect of outsourcing services on harvest loss rates. c_1' captures the direct effect of outsourcing services on harvest loss rates. z_i represents the associated covariates, with its corresponding estimator c_i without mediator and b_i with mediator. Based on our analysis and existing research (Basavaraja et al. 2007; Qu et al. 2020), covariates are introduced in the "Covariates" (see Sect. 5.2.1.4). c_0 and b_0 are the intercepts, while e_1 and e_2 are the error terms, which are normally distributed and independent of explanatory variables.

Equation (5.7) illustrates Fig. 5.3a, which shows the total effect of harvest outsourcing services on rice harvest loss rates. Equation (5.8) describes the direct effect of harvest outsourcing services on rice harvest loss rates adjusted by work attitudes, as shown in Fig. 5.3b. As with Chap. 4, Logit regression is used for the mediator component to analyze the effect of outsourcing services on operators' work attitudes as shown in Eq. (5.9).

$$\ln\left(\frac{\mathrm{Prob}(\mathrm{WA}=1)}{\mathrm{Prob}(\mathrm{WA}=0)}\right) = a_0 + a_1(\mathrm{OS}) + \sum_{i=2}^{k} a_i z_{1i} \tag{5.9}$$

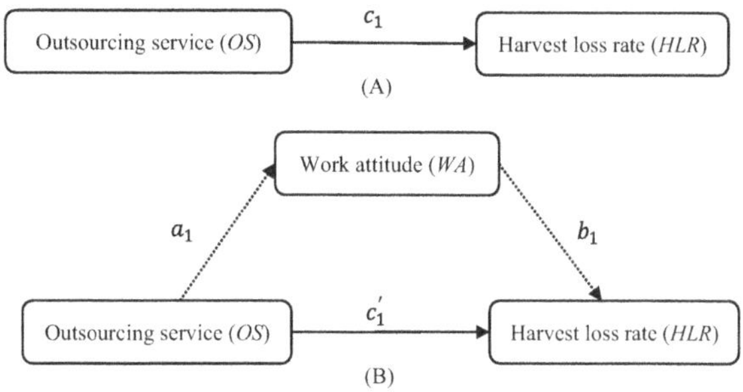

Fig. 5.3 Mediation model

where a_1 captures the effect of outsourcing services on work attitudes. z_{1i} represents the associated covariates (see Sect. 5.2.1.4).

5.2.2.3 Three-Step Method

Since Eqs. (5.7) and (5.8) are Tobit models and Eq. (5.9) is a Logit model, neither the product of coefficients method ($a_1 \times b_1$) nor the difference in coefficients method ($c_1 - c_1'$) is suitable for mediation estimation (Li et al. 2007). We follow the practice of Li and Zhao (2018) and use the three-step method to test the mediation effects. Mediation effects are thought to be present if the marginal effect of outsourcing services is significant in Eq. (5.7), but its magnitude and/or significance level are reduced in Eq. (5.8) (relative to Eq. (5.7)), and it is significant in Eq. (5.9) (path a_1). The logic of mediation analysis is that outsourcing services affect the mediator (work attitude), which in turn affect the harvest loss rate. We compare the changes in the marginal effects of outsourcing services on harvest loss rates, with and without the mediator (i.e., in Eqs. (5.7) and (5.8)). The difference between the marginal effect of outsourcing services with and without the mediator indicates the magnitude of the effect of outsourcing services on the harvest loss rates, as reflected through the mediator.

5.2.2.4 The Adjusted Product of Coefficients Method

Although some farmers estimated the rice harvest loss rate to be 0% (approximately 6% of the farmers), it is currently impossible to avoid harvest losses entirely with existing technology. Therefore, we winsorize the sample with a harvest loss of 0. Specifically, if the harvest loss per unit area was less than the value at the 10th percentile, it is replaced by the value at the 10th percentile. Then Ordinary Least Squares (OLS) is used for the estimation of outsourcing services on harvest loss rates with and without work attitudes as shown in Eqs. (5.10) and (5.11).

$$\mathrm{HLR} = c_0 + c_1(\mathrm{OS}) + \sum_{i=2}^{k} c_i z_i + e_1 \tag{5.10}$$

$$\mathrm{HLR} = b_0 + b_1(\mathrm{WA}) + c_1'(\mathrm{OS}) + \sum_{i=2}^{k} b_i z_i + e_2 \tag{5.11}$$

While the three-step method is very straightforward, it receives much critique (Mackinnon 2008). Some researchers contend that the three-step method cannot test the magnitude and significance of the mediation effects, whereas the product of coefficients method based on path analysis can (MacKinnon and Fairchild 2009). Considering that Eq. (5.9) is Logit model, here we use the adjusted product of coefficients method. Iacobucci (2012) emphasized that this method can be used in any combination of OLS regressions and Logit regressions. Firstly, the coefficients (a_1 and b_1) of Logit or OLS regression should be adjusted by Eqs. (5.12) and (5.13) to obtain the adjusted coefficients Z_{a_1} and Z_{b_1}, in which s_{a_1} and s_{b_1} represent the standard errors of a_1 and b_1, respectively. Then the mediation effect can be expressed as $Z_{a_1 \times b_1}$ in Eq. (5.14). The Z-test (i.e., the so-called Sobel test) is used to test its significance, that is, if the absolute value of $Z_{\text{mediation}}$ in Eq. (5.15) is greater than 1.96 (for two-tailed tests and $\alpha = 0.05$), then the mediation effect is significant (Iacobucci 2012).

$$Z_{a_1} = \frac{a_1}{s_{a_1}} \tag{5.12}$$

$$Z_{b_1} = \frac{b_1}{s_{b_1}} \tag{5.13}$$

$$Z_{a_1 \times b_1} = Z_{a_1} \times Z_{b_1} \tag{5.14}$$

$$Z_{\text{mediation}} = \frac{Z_{a_1 \times b_1}}{\mathrm{SE}\left(Z_{a_1 \times b_1}\right)} = \frac{Z_{a_1} \times Z_{b_1}}{\sqrt{Z_{a_1}^2 + Z_{b_1}^2 + 1}} \tag{5.15}$$

However, the two-tailed *p*-value, based on the normal distribution assumption of $Z_{a_1 \times b_1}$ required by the Sobel test, has been seriously questioned (Preacher and Hayes 2004). To improve this, MacKinnon and Cox (2012) recommended using the distribution of the product, while Feinberg (2012) suggested using Bayesian techniques to obtain more accurate confidence limits and statistical tests.

In summary, the mediation effects are tested by two methods in this study. First, the three-step method is used to test the existence of the mediation effects when the effects of outsourcing services on harvest loss rates with and without work attitudes are estimated by Tobit models (Eqs. 5.7 and 5.8), using the sample before winsorizing. This could be realized by STATA 15.0. Second, the adjusted product of coefficients method suggested by Iacobucci (2012) is applied when the effects of outsourcing services on harvest loss rates with and without work attitudes are estimated by OLS models (Eqs. 5.10 and 5.11), using the sample after winsorizing. In the second method, the Sobel test, the distribution of the product method, and the Markov Chain Monte Carlo (MCMC) method are all used to test the significance of the mediation effect. This can be performed by STATA 15.0 and *R*.

5.2.3 Issues of Endogeneity

The endogeneity of outsourcing choice is a focus in studying the effect of outsourcing services on agricultural production. The endogeneity of outsourcing service choice is caused by omitted variable bias resulting from unobserved factors that affect both harvest losses and outsourcing choice. The main factors affecting the decision of outsourcing are terrain, plot location, and market availability, among which the terrain is the essential factor that also affects harvest losses. Fortunately, our dataset contains this crucial information about terrain to avoid the endogeneity caused by unobserved factors. Some studies thought that the reverse causality is also a source of endogeneity in outsourcing choice. We do not support this view. Lu and Du (2020) also disagree with this point when estimating the effect of outsourcing on crop yield. They argued that outsourcing choices are made before the yields are realized and that the crop yields cannot have an influence on outsourcing choices in the current season. We hold the same view that outsourcing choices are made before harvest losses and that the latter cannot influence the former. During the harvest season, the demand for outsourcing services is usually high. The main factors influencing outsourcing decisions are household labor, terrain conditions, and market supply, rather than harvest loss. In existing studies, the decision to outsource harvesting has never been studied with harvest losses as a possible influencing factor.

One might be concerned that work attitude is endogenous in Eqs. (5.8) and (5.11) due to reverse causality. Service providers are more likely to implement moral hazards when working on plots with potentially greater harvest losses. Because it is difficult to distinguish whether the loss is due to the poor conditions of the plot itself (such as rough terrain) or due to careless work. To address this, we employ instrumental variable approach and use IV-Tobit and 2SLS to estimate Eqs. (5.8) and (5.11) (Angrist and Pischke 2008).

The instrumental variable is the distance from the homestead to the nearest paved road (Htor). The latter must be strongly correlated with the work attitude—the so-called inclusion restriction. We assume that the farther distance between the homestead and the nearest paved road, which means a longer journey on unpaved roads, translates to more serious work attitudes, since driving on unpaved roads requires caution. In addition, economically underprivileged farmers and elder farmers are more likely to live in remote areas (Dercon et al. 2009; Fan et al. 2000). They typically dislike losses, and their attitudes toward valuing grain can affect the work attitude of operators.

In addition, the instrumental variable must be related to the harvest loss rates only through its effect on work attitudes—the so-called exclusion restriction. Intuitively, the exclusion restriction is likely to hold. Because the distance between the homestead and the nearest paved road depends on the locations of these two points, the homestead and the paved road, which are planned by the village collectives. It offers an exogenous source of variation that affects the willingness to provide services (and thus influencing service providers' work attitudes) but is unlikely to affect harvest losses. Therefore, although the exclusion restriction is not testable, we consider Htor to be a valid instrumental variable.

5.3 Results and Discussion

5.3.1 Variable Description Statistics

Table 5.1 lists the summary and definition of variables used in the empirical analysis. The average harvest loss rates before and after winsorizing were remarkably close, at 3.65% and 3.69%, respectively. It means that about 7.7 million tons of rice, which required 1 million hectares of farmland to produce, was left in the field and failed to be harvested.

More than half of the farmers purchased outsourcing services. Approximately 50% of the farmers used combine harvesters, mechanical winnowing, and mechanical field transport, among which the proportion of using combine harvesters was the lowest and that of using mechanical field transport was the highest.

5.3.2 Mediation Test Results of Three-Step Method

To investigate the path by which harvest outsourcing services affect rice harvest loss rates through the work attitudes of operators, we conduct the three-step method.

Firstly, we conduct a Tobit regression in Eq. (5.7) to investigate the effect of harvest outsourcing services on rice harvest loss rates without the mediator. The estimation results are presented in column (1) of Table 5.2. Without the mediator, the marginal effect of outsourcing services (OS) on rice harvest loss rates is 0.393, which is significant at the $\alpha = 0.1$ level. It indicates that the purchase of harvest outsourcing services increases rice harvest loss rates.

Then the Logit regression is estimated to study the effect of outsourcing services on work attitudes as shown in Eq. (5.9). The estimation results in column (3) of Table 5.2 show that the marginal effect of outsourcing services (OS) on work attitudes is −0.342, which is significant at the $\alpha = 0.01$ level. It means that purchasing harvest outsourcing services reduces the probability of operators' (i.e., the service provider) serious work attitudes.

Before moving to the effect of harvest outsourcing services on rice harvest loss rates with work attitudes, we need to address the risk of potential endogeneity caused

Table 5.1 Summary and definition of variables (1106 farms)

Variable	Definition	Mean	Std. Dev
HLR	Harvest loss rate (%)	3.65	3.60
HLR	Harvest loss rate (%) (after the winsorizing)	3.69	3.56
Core independent variables			
OS	Dummy = 1 if farmer bought outsourcing service, 0 otherwise	0.59	0.49
Mediator			
WA	Dummy = 1 if operator's harvest attitude was serious, 0 otherwise	0.23	0.42
Instrumental variable			
Htor	Distance from homestead to the nearest paved road	0.34	0.76
Production and harvesting variables			
Com	Dummy = 1 if combine harvesting, 0 if segmented harvesting	0.46	0.50
Win	Dummy = 1 if machinery winnowing, 0 otherwise	0.53	0.50
Tra	Dummy = 1 if machinery field transport, 0 otherwise	0.60	0.49
Wea	Dummy = 1 if bad weather when harvesting, 0 if normal weather	0.16	0.37
Pest	No pest = 1, slight pests = 2, general or serious pests = 3	1.84	0.76
Area	Rice planting area (ha)	0.33	0.34
Yield	Yield (quintal/ ha)	80.45	25.71
Flat	Dummy = 1 if plot terrain is flat, 0 otherwise	0.75	0.43
Dis	Distance from the field to storage locations (km)	0.65	0.74
Labor	Dummy = 1 if farmer reported a lack of labor, 0 otherwise	0.28	0.45
Sav	Dummy = 1 if farmer picked up rice left in field, 0 otherwise	0.16	0.37
Mat	Dummy = 1 if rice was mature when harvesting, 0 otherwise	0.95	0.23
Price	Sale price of rice (CNY/kg)	2.98	0.34
Household and individual variables			
Gen	Gender of household head (male = 1, female = 0)	0.84	0.36
Age	Age of household head	54.12	10.63
Edu	School years of household head (years)	7.01	2.66
Train	Dummy = 1 if household head had agricultural training, 0 otherwise	0.09	0.29
Tinc	Household income (ten thousand CNY)	7.07	5.45
Rincs	Rice income as a percentage of total income (%)	15.80	18.16
N	Number of samples	1106	

Notes (1) The highest correlation coefficient between independent variables is no more than 0.8 and the VIF-value is less than 10, which means there is no multicollinearity
Data source Author's calculation based on the survey

Table 5.2 Mediation effect test: three-step method (1106 farms)

Variable	Rice harvest loss rate (%)				Work attitude	
	(1) Tobit (marginal effect)		(2) IV-Tobit (marginal effect)		(3) Logit (marginal effect)	
Core independent variables						
OS	0.393*	(0.21)	0.132	(0.60)	−0.342***	(0.04)
Mediator						
WA			−0.934	(1.85)		
Production and harvesting variables						
Com	0.046	(0.19)	0.140	(0.27)	0.144***	(0.04)
Win	0.679***	(0.14)	0.636***	(0.17)		
Tra	−0.779***	(0.16)	−0.874***	(0.19)		
Wea	0.914***	(0.24)	0.764***	(0.28)	−0.094**	(0.04)
Pest = 2	0.727***	(0.15)	0.579**	(0.27)	−0.108***	(0.03)
Pest = 3	1.685***	(0.20)	1.631***	(0.30)	−0.115***	(0.03)
Area	−1.375***	(0.29)	−1.260**	(0.52)	0.221***	(0.05)
Yield	−0.003	(0.00)	−0.004	(0.00)		
Flat	−0.046	(0.18)	0.034	(0.25)	0.105***	(0.03)
Dis	−0.139	(0.09)	−0.071	(0.09)		
Labor	0.475***	(0.15)	0.437***	(0.16)	−0.023	(0.03)
Sav	0.515***	(0.17)	0.637***	(0.20)	0.057*	(0.03)
Mat	0.137	(0.29)	0.166	(0.31)		
Price	−0.163	(0.22)	0.132	(0.24)	0.030	(0.04)
Household and individual variables						
Gen	−0.035	(0.18)	0.010	(0.18)	0.014	(0.03)
Age	0.015*	(0.01)	0.017**	(0.01)	0.002*	(0.00)
Edu	−0.004	(0.03)	0.002	(0.03)	0.006	(0.00)
Train	0.446*	(0.24)	0.420*	(0.24)	−0.025	(0.05)
Tinc	0.017	(0.01)	0.016	(0.02)	−0.005*	(0.00)
Rincs	−0.003	(0.01)	−0.003	(0.01)	−0.002**	(0.00)
Htor					0.048***	(0.01)
Advantageous region	Yes		Yes		Yes	
N	1106		1106		1106	
Pseudo R^2	0.046				0.147	

Notes (1) ***, **, and * indicate 1, 5, and 10% significance levels, respectively. (2) Robust standard errors statistics are in parentheses

Data source Author's calculation based on the survey

by the reverse causality between harvest loss rates and work attitudes. Therefore, we apply IV-Tobit for the estimation of Eq. (5.8). As a statistical test, the Wald tests of exogeneity in Table 5.3 are 17.67 and 18.01 without and with covariates, respectively. They are both significant, which implies that the work attitude is endogenous. It is necessary to address the endogeneity of work attitudes. The coefficients of the instrument variable on work attitudes without and with covariates are 0.077 and 0.065, respectively, which are significant at the $\alpha = 0.01$ level, implying that the instrument variable is strongly correlated the work attitudes. Moreover, the Kleibergen–Paap F statistic (Kleibergen and Paap 2006) is 15.920 without the covariates and 11.506 with the covariates, both of which are larger than the rule of thumb of 10.Therefore, the instrument variable could not be regarded as weak.

Moreover, we explore the possibility that the instrumental variable satisfies the exclusion restrictions by testing whether it correlates with some possible covariates—the land terrain and rice planting area. Previous study has shown that the more the instrumental variable correlates with covariates, the higher the possibility of violating the exclusion restrictions (Wuepper et al. 2018). Columns (3) and (4) in Table 5.3 give that the instrumental variable is not significantly correlated with these two variables.

Column (2) in Table 5.2 contains the marginal effect of harvest outsourcing services on rice harvest loss rates with the mediator after excluding the potential

Table 5.3 Tests of instrumental variable

	Work attitude		(3) Flat	(4) Area
	First stage (IV-Tobit)			
	(1) Without covariates	(2) With covariates		
Htor	0.077*** (0.02)	0.065*** (0.02)	−0.053 (0.89)	−0.009 (0.01)
Constant	0.251** (0.04)	−0.100 (0.19)	1.728 (0.57)	0.263 (0.06)
Production and production control	No	Yes	No	No
Household and individual control	No	Yes	Yes	Yes
Advantageous region	Yes	Yes	Yes	Yes
Kleibergen–Paap F statistic	15.920	11.506		
Wald test of exogeneity	17.67***	18.01***		
Observations	1106	1106	1106	1106

Notes (1) ***, **, and * indicate 1, 5, and 10% significance levels, respectively. (2) Robust standard errors statistics are in parentheses. (3) Logit regression and ordinary least squares are used for estimation of column (3) and (4), respectively. (4) The results of other variables can be found in column (1) and (2) of Table A4 in Appendices

Data source Author's calculation based on the survey

endogeneity of work attitudes. In the presence of mediator, the marginal effect of outsourcing services (OS) on harvest loss rates is 0.132, which remains positive but insignificant. After the inclusion of the mediator, the marginal effect of outsourcing services on harvest loss rates becomes insignificant and smaller than before. Both the magnitude and significance level of the marginal effect of outsourcing services decrease. Therefore, the three-step method suggests that harvest outsourcing services have an impact on rice harvest loss rates through their impact on work attitudes.

Furthermore, we will briefly discuss other factors that affect rice harvest loss rates.[5] Column (2) in Table 5.2 gives that the marginal effect of mechanical winnowing (Win) is positive, which means that winnowing by machine will cause higher losses compared to manual operation. The machine is a complex system that is influenced by maintenance and operation. Mechanical winnowing is affected by feed rate and air speed. If the feed rate is too large or air speed is too fast, grain losses will increase (Ei-Awady et al. 2009). The marginal effect of mechanical field transport (Tra) is negative, which indicates that mechanical transport can reduce harvest loss rates. Apparently, mechanical transport is faster than that human or animal power, which can reduce the losses during field transport. The marginal effects of bad weather (Wea) and pests (Pest) are positive. This is consistent with our intuition that bad weather and pests will unsurprisingly increase harvest loss rates. The marginal effect of planting area (Area) is negative, indicating that increased planting area will reduce rice harvest loss rates. The importance of rice cultivation increases as the area increases. Farmers who have planted a larger area of rice will be more willing to invest in field management, such as pest and disease control, which helps reduce harvest losses. Moreover, harvesting is a labor-intensive operation. As we can see, the marginal effect of labor (Labor) is positive, implying that the shortage of labor leads to harvest loss rates.

5.3.3 Mediation Test Results of the Adjusted Product of Coefficients Method

To further test the significance of the mediation effect, we apply the adjusted product of coefficients method. In this test, we only need to estimate Eqs. (5.9) and (5.11). Table 5.4 gives the coefficients that we need for the test—path a_1 and path b_1. Column (1) in Table 5.4 gives the coefficient of work attitudes (WA) on harvest loss rates in Eq. (5.11), which is negative and significant (−5.540). Column (2) reports the coefficient of outsourcing services (OS) on work attitudes in Eq. (5.9), which is also negative and significant (−2.295). The last four rows present the mediation effects and the corresponding significance tests by the Sobel test, the distribution of the product test, and the MCMC test. The Sobel test gives a mediation effect of 20.085, with the corresponding $Z_{\text{mediation}}$ of 2.527. Therefore, the mediation effect is

[5] The results without mediator are only for the purpose of mediation model analysis. To avoid estimation bias due to omitted variables, the results with the mediator are discussed here.

significant at the $\alpha = 0.05$ level since the value of $Z_{\text{mediation}}$ is greater than 1.96 (for the two-tailed test with $\alpha = 0.05$).

The Sobel Test is based on a large sample size and normal distribution assumption, in which $Z_{a_1 \times b_1}$ does not always have a normal distribution, even though both Z_{a_1} and Z_{b_1} are normally distributed (MacKinnon et al. 2004). To obtain more accurate confidence limits and statistical tests, the distribution of product method suggested by MacKinnon and Cox (2012) and the MCMC method suggested by Feinberg (2012) are used. The distribution of product method shows that the mediation effect is 12.714, which is smaller than that shown by the Sobel test. The corresponding 95% confidence intervals are (3.409, 23.170), in which 0 is not included. It means that outsourcing services significantly affect rice harvest loss rates as mediated by operators' work attitudes at $\alpha = 0.05$ level. The MCMC gives the mediation effect of 12.709 and its corresponding 95% confidence intervals are (3.374, 23.167). The results from the MCMC method are similar to those from the distribution of product method. Therefore, work attitudes significantly mediate the effect of harvest outsourcing services on rice harvest losses.

Table 5.4 Mediation effect test: the adjusted product of coefficients method (1106 farms)

	Rice harvest loss rate (%)	Work attitude
	(1) Harvest loss rate with mediators (2SLS)	(2) Path *a* (logit)
OS		-2.295^{***} (0.31)
WA	-5.540^{***} (2.04)	
Production and harvesting control	Yes	Yes
Household and individual control	Yes	Yes
Advantageous region	Yes	Yes
Kleibergen–Paap F	11.506	
Endogeneity test	$p = 0.013$	
Observation	1106	1106
Mediation effect test	*Mediation effect*	$Z_{mediation}$/*95% CIs*
The Sobel test	20.085	2.527
The distribution of the product	12.714	[3.409, 23.170]
Markov Chain Monte Carlo	12.709	[3.374, 23.167]

Notes (1) ***, **, and * indicate 1, 5, and 10% significance levels, respectively. (2) Robust standard errors statistics are in parentheses. (3) CIs means confidence intervals. (4) The results of other variables can be found in column (2) and (3) of Table A1 in Appendices. (5) The first stage results can be found in Table 5.3

Data source Author's calculation based on the survey

5.4 Robustness Analysis

5.4.1 Mediation Analysis on Large-Scale Farms

The mediation effect has been proven by the three-step method and the adjusted product of coefficients method. However, the positive effect of outsourcing services without mediator has a low significance level on harvest loss rates—only 1% significance level in Tobit regression (see column (1) in Table 5.2) and insignificant in OLS regression (see column (1) of Table A1 in Appendices). In China, when the reaping is in progress, farmers usually wait on-site for the subsequent work (such as threshing and bagging), which plays a supervisory role, thereby reducing the possibility of moral hazards to some extent. When planting area increases, such supervision becomes difficult to enforce due to management difficulties. Moreover, as the discussion in Chaps. 3 and 4, the implementation of certain moral hazard behaviors (such as increased forward speed of machine) requires certain conditions—a relatively large planting area. Therefore, the inclusion of small-scale farms may be the reason for the low significance level for outsourcing services in above regressions. Consequently, we exclude small-scale farms and conduct the robustness test on large-scale farms. The large-scale farms are the same as those in Chap. 4.

5.4.1.1 Mediation Test Results of Three-Step Method

Table 5.5 gives the results of three-step method for large-scale farms. Without mediator, the marginal effect of outsourcing services (OS) is 0.795, which is significant at $\alpha = 0.01$ level. Both the magnitude and significance level increase after excluding the small-scale farms (compared to Table 5.2).

Furthermore, the coefficient for harvest outsourcing services becomes significantly positive (see column (1) of Table A2 in Appendices, compared to column (1) of Table A1). This confirms above analysis that moral hazard is more pronounced in relatively large-scale farms.

Next, we proceed to the three-step mediation analysis. In column (2), both the F statistic and Wald test of exogeneity indicate the reliable of IV method. With the work attitude, the marginal effect of outsourcing services (OS) on harvest loss rates is 0.722 and significant at $\alpha = 0.05$ level. Both the magnitude and significance level of the marginal effect of outsourcing services decrease after including the mediator. Meanwhile, the marginal effect of harvest outsourcing services (OS) on work attitude is significantly negative in column (3). Therefore, based the three-step method, harvest outsourcing services have an impact on harvest loss rates through its impact on work attitudes on large-scale farms.

Table 5.5 Mediation effect test: three-step method (large-scale farms)

Variable	Rice harvest loss rate (%)				Work attitude	
	(1) Tobit (marginal effect)		(2) IV-Tobit (marginal effect)		(3) Logit (marginal effect)	
Core independent variables						
OS	0.795***	(0.26)	0.722*	(0.41)	−0.338***	(0.06)
Mediator						
WA			−0.391	(0.94)		
Production and harvesting variables						
Com	−0.201	(0.21)	−0.110	(0.28)	0.179***	(0.05)
Win	0.509***	(0.18)	0.404**	(0.20)		
Tra	−0.365*	(0.21)	−0.429**	(0.21)		
Wea	0.719**	(0.33)	0.614*	(0.35)	0.073	(0.06)
Pest = 2	0.606***	(0.18)	0.445**	(0.21)	−0.095**	(0.04)
Pest = 3	1.130***	(0.24)	1.015***	(0.28)	−0.150***	(0.04)
Area	−0.618***	(0.24)	−0.601**	(0.29)	0.125***	(0.05)
Yield	−0.004	(0.00)	−0.005	(0.00)		
Flat	0.260	(0.21)	0.298	(0.21)	−0.028	(0.04)
Dis	−0.017	(0.08)	−0.043	(0.08)		
Labor	0.455***	(0.18)	0.425**	(0.19)	−0.066	(0.05)
Sav	0.329*	(0.19)	0.442**	(0.19)	0.028	(0.05)
Mat	0.002	(0.31)	0.059	(0.32)		
Price	−0.565	(0.47)	−0.400	(0.58)	−0.290***	(0.11)
Household and individual variables						
Gen	−0.042	(0.22)	−0.019	(0.22)	0.054	(0.05)
Age	0.008	(0.01)	0.009	(0.01)	0.002	(0.00)
Edu	−0.040	(0.04)	−0.040	(0.04)	0.001	(0.01)
Train	0.162	(0.28)	0.145	(0.28)	0.011	(0.07)
Tinc	0.040**	(0.02)	0.047***	(0.02)	−0.003	(0.00)
Rincs	−0.008	(0.01)	−0.007	(0.01)	−0.001	(0.00)
Htor					0.104***	(0.03)
Advantageous region	Yes		Yes		Yes	
Pseudo R^2	0.041				0.157	
Kleibergen–Paap *F*			14.535			

(continued)

Table 5.5 (continued)

Variable	Rice harvest loss rate (%)		Work attitude
	(1) Tobit (marginal effect)	(2) IV-Tobit (marginal effect)	(3) Logit (marginal effect)
Wald test of exogeneity		16.12***	
N	532	532	532

Notes (1) ***, **, and * indicate 1, 5, and 10% significance levels, respectively. (2) Robust standard errors statistics are in parentheses. (3) The first stage results of IV-Tobit can be found in column (3) of Table A4 in Appendices

Data source Author's calculation based on the survey

5.4.1.2 Mediation Test Results of the Adjusted Product of Coefficients Method

Table 5.6 gives the results of the adjusted product of coefficients method for large-scale farms. The mediation effect given by the Sobel test is 12.331 and its corresponding $Z_{\text{mediation}}$ is 2.101, which is greater than 1.96 (for the two-tailed test with $\alpha = 0.05$). The distribution of the product and MCMC method indicate the mediation effects of 6.613 and 6.604, respectively. The 95% confidence interval for the distribution of the product method is (0.991, 13.338), while the MCMC method gives the confidence interval of (1.003, 13.352). Both of these 95% confidence intervals exclude 0, which means that outsourcing services significantly affect harvest loss rates on large-scale farms at the $\alpha = 0.05$ level, mediated by operators' work attitudes.

5.4.2 *Mediation Analysis on Large-Scale Farms Using Combines*

As indicated before, rice harvesting can be done by combine harvesting and segmented harvesting. Combine harvesters are more advanced than the machines used for segmented harvesting, but they are also more expensive, hence the cost of purchasing combine harvest outsourcing services is higher (Poungchompu and Chantanop 2016). In Chap. 4, we find that using combine harvesters can mitigate service providers' moral hazards. Therefore, we assume that the higher price of combine harvest outsourcing services—which could be regarded as income incentives for service providers—will reduce the possibility of moral hazard and its effect on rice harvest losses. Therefore, we further conduct the mediation analysis on large-scale farms using combine harvesting.

Table 5.6 Mediation effect test: the adjusted product of coefficients method (large-scale farms)

	Rice harvest loss rate (%)	Work attitude
	(1) Harvest loss rate with mediators (2SLS)	(2) Path *a* (logit)
OS		-2.371^{***} (0.45)
WA	-2.789^{***} (1.20)	
Production and harvesting control	Yes	Yes
Household and individual control	Yes	Yes
Advantageous region	Yes	Yes
Kleibergen–Paap *F*	14.535	
Endogeneity test	$p = 0.028$	
Observation	532	532
Mediation effect	*Mediation effect*	$Z_{mediation}$/95% *CIs*
The Sobel test	12.331	2.101
The distribution of the product	6.613	[0.991, 13.338]
Markov Chain Monte Carlo	6.604	[1.003, 13.352]

Notes (1) ***, **, and * indicate 1, 5, and 10% significance levels, respectively. (2) Robust standard errors statistics are in parentheses. (3) CIs means confidence intervals. (4) The results of other variables can be found in column (2) and (3) of Table A2 in Appendices. (5) The first stage results can be found in column (3) of Table A4 in Appendices
Data source Author's calculation based on the survey

5.4.2.1 Mediation Test Results of Three-Step Method

Table 5.7 gives the results of three-step method. Firstly, without work attitude, the marginal effect of outsourcing services (OS) on rice harvest loss rates is 0.342, which is insignificant as shown in column (1). It means that using harvest outsourcing services has no effect on rice harvest loss rates. Nevertheless, we cannot stop the test. Because the presence of a mediation effect does not necessarily require the total effect to be significant (Preacher and Hayes 2004). In column (3), the marginal effect of outsourcing services (OS) on work attitudes is −0.146, which is still insignificant. It means that harvest outsourcing services have no effect on work attitudes. These results fail the three-step method. Therefore, harvest outsourcing services have no effect on rice harvest loss rates through operators' work attitudes.

5.4.2.2 Mediation Test Results of the Adjusted Product of Coefficients Method

Table 5.8 gives the results of the adjusted product of coefficients method. In column (1), both the *F* statistic and *p*-value for the endogeneity test indicate that the estimation results of 2SLS are reliable. The mediation effect given by the Sobel test is 2.734. Its corresponding $Z_{\text{mediation}}$ is 1.005, which is smaller than 1.96 (for the two-tailed test

Table 5.7 Mediation effect test: three-step method (large-scale farms using combines)

Variable	Rice harvest loss rate (%)				Work attitude	
	(1) Tobit (marginal effect)		(2) IV-Tobit (marginal effect)		(3) Logit (marginal effect)	
Core independent variables						
OS	0.342	(0.32)	0.185	(0.38)	−0.146	(0.12)
Mediator						
WA			−0.673	(0.70)		
Production and harvesting variables						
Win	0.793***	(0.25)	0.760***	(0.26)		
Tra	−0.421	(0.27)	−0.417	(0.27)		
Wea	0.107	(0.36)	0.005	(0.38)	0.118*	(0.07)
Pest = 2	0.529**	(0.23)	0.394*	(0.23)	0.027	(0.06)
Pest = 3	0.626**	(0.30)	0.411	(0.30)	−0.186***	(0.07)
Area	−0.493*	(0.26)	−0.356	(0.29)	0.156**	(0.07)
Yield	−0.001	(0.00)	−0.002	(0.00)		
Flat	0.458	(0.30)	0.598*	(0.31)	−0.039	(0.07)
Dis	0.036	(0.13)	−0.011	(0.13)		
Labor	0.060	(0.21)	−0.067	(0.24)	−0.195**	(0.08)
Sav	0.076	(0.25)	0.136	(0.25)	−0.058	(0.08)
Mat	−0.482	(0.33)	−0.424	(0.33)		
Price	−0.530	(0.57)	−0.017	(0.62)	−0.170	(0.17)
Household and individual variables						
Gen	0.284	(0.25)	0.288	(0.24)	0.009	(0.06)
Age	0.008	(0.01)	0.006	(0.01)	−0.003	(0.00)
Edu	−0.048	(0.04)	−0.052	(0.04)	−0.002	(0.01)
Train	0.223	(0.34)	0.187	(0.35)	−0.271	(0.18)
Tinc	0.023	(0.02)	0.031	(0.02)	−0.002	(0.00)
Rincs	−0.016***	(0.01)	−0.015***	(0.01)	−0.002	(0.00)
Htor					0.145***	(0.03)
Advantageous region	Yes		Yes		Yes	
Pseudo R^2	0.058				0.220	
Kleibergen–Paap *F*			17.108			

(continued)

Table 5.7 (continued)

Variable	Rice harvest loss rate (%)		Work attitude
	(1) Tobit (marginal effect)	(2) IV-Tobit (marginal effect)	(3) Logit (marginal effect)
Wald test of exogeneity		15.83***	
N	302	302	302

Notes (1) Robust standard errors are reported in parentheses. (2) Significance levels are as follows: *** = 1%, ** = 5%, and * = 10%. (3) The first stage results of IV-Tobit can be found in column (4) of Table A4 in Appendices
Data source Author's calculation based on the survey

with $\alpha = 0.05$). It is even smaller than 1.65 (for the two-tailed test with $\alpha = 0.1$). Therefore, the mediation effect given by the Sobel test is not significant. This means that in large-scale farms using combine harvesting, the mediation effect of work attitudes does not exist.

Next, the more effective test of the distribution of product method and the MCMC method will be used to further test the significance of mediation effect. The 95%

Table 5.8 Mediation effect test: the adjusted product of coefficients method (large-scale farms using combines)

	Rice harvest loss rate (%)	Work attitude
	(1) Harvest loss rate with mediators (2SLS)	(2) Path *a* (Logit)
OS		−0.984 (0.79)
WA	−2.005** (0.91)	
Production and harvesting control	Yes	Yes
Household and individual control	Yes	Yes
Advantageous region	Yes	Yes
Kleibergen–Paap *F*	17.108	
Endogeneity test	$p = 0.085$	
Observation	302	302
Mediation effect test	*Mediation effect*	$Z_{mediation}$/*95% CIs*
The Sobel test	2.734	1.005
The distribution of the product	1.973	[−1.149, 6.568]
Markov Chain Monte Carlo	1.972	[−1.154, 6.527]

Notes (1) ***, **, and * indicate 1, 5, and 10% significance levels, respectively. (2) Robust standard errors statistics are in parentheses. (3) CIs means confidence intervals. (4) The results of other variables can be found in column (2) and (3) of Table A3 in Appendices. (5) The first stage results can be found in column (4) of Table A4 in Appendices
Data source Author's calculation based on the survey

confidence interval given by the distribution of the product method is (−1.149, 6.568), while the 95% confidence interval given by the MCMC method is (−1.154, 6.527). The value of 0 is included in these two 95% confidence intervals, which means that the mediation effects are not significant. In fact, the 90% confidence intervals also include 0. Therefore, harvest outsourcing services have no effect on rice harvest loss rates through operators' work attitudes, suggesting that there is no moral hazard in outsourcing services if large-scale farmers choose combine harvesting. The high income from combine harvesting may be the reason for reducing moral hazards.

5.4.3 Regional Control of Rice Cropping Regionalization

As in Chap. 4, the rice advantageous regional dummy variables are changed to rice cropping regional dummy variables for robustness test. The specific classification of rice cropping regions is the same as that in Chap. 4. Table 5.9 gives the results of the adjusted product of coefficients method for 1106 sample using rice cropping regional controls. First, both the F statistic and p-value for the endogeneity test indicate that the 2SLS estimates are reliable. The mediation effect given by the Sobel test is 20.204. Its corresponding $Z_{\text{mediation}}$ of 2.715 is greater than 1.96 (for the two-tailed test with $\alpha = 0.05$), which implies that the mediation effect is significant at the $\alpha = 0.05$ level.

Table 5.9 Mediation effect test: the adjusted product of coefficients method (rice cropping regional control)

	Rice harvest loss rate (%)	Work attitude
	(1) Harvest loss rate with mediators (2SLS)	(2) Path *a* (Logit)
OS		-2.176^{***} (0.32)
WA	-5.944^{***} (1.98)	
Production and harvesting control	Yes	Yes
Household and individual control	Yes	Yes
Rice cropping region	Yes	Yes
Kleibergen–Paap *F*	13.821	
Endogeneity test	$p = 0.003$	
Observation	1106	1106
Mediation effect test	*Mediation effect*	$Z_{mediation}$/95% CIs
The Sobel test	20.204	2.715
The distribution of the product	12.934	[4.261, 22.945]
Markov Chain Monte Carlo	12.960	[4.329, 22.972]

Notes (1) ***, **, and * indicate 1, 5, and 10% significance levels, respectively. (2) Robust standard errors statistics are in parentheses. (3) CIs means confidence intervals
Data source Author's calculation based on the survey

The more accurate confidence limits are given by the distribution of product method and the MCMC method. The distribution of product method indicates a mediation effect of 12.934, with a corresponding 95% confidence interval of (4.261, 22.945), which excludes 0. The results from the MCMC method are similar to those from the distribution of product method. The mediation effect given by the MCMC method is 12.960 and its corresponding 95% confidence interval is (4.329, 22.972). 0 is also not included in the confidence interval. Therefore, work attitudes significantly mediate the effect of outsourcing services on harvest loss rates.

Moreover, using rice cropping regional dummy variables, the significant mediation effect of work attitudes is found for large-scale farms and the insignificant mediation effect of work attitudes is found for large-scale farms using combine harvesters. These results are also similar to those using rice advantageous regional dummy variables.

5.5 Summary

To examine whether the negative effect of harvest outsourcing services on operators' work attitudes leads to an increase in rice harvest losses, this chapter uses mediation analysis models to test the hypothesis that moral hazards—measured by whether harvest operators' work attitudes are serious— mediate the effect of harvest outsourcing services on rice harvest losses.

In harvest outsourcing services, inconsistent goals and information asymmetry between farmers and service providers may lead to moral hazards. Here, service providers, as the more informed parties, have motivations to maximize their own interests by taking actions such as increased forward speed of machines, which will cause carelessness and higher harvest losses. This is examined using mediation analysis models that demonstrate the impact of harvest outsourcing services on the mediator (the work attitude of harvest operator) in turn leading to an increase in rice harvest losses. After controlling for the potential endogeneity of work attitudes, the mediation effect of moral hazards is preliminarily tested by the three-step method. The marginal effect of outsourcing services on harvest loss rates in Tobit regression is significant and positive, and it becomes insignificant in IV-Tobit regression after the mediator is included as an independent variable. Meanwhile, the marginal effect of outsourcing services on work attitudes is significant and negative. Then, the adjusted product of coefficients method with the Sobel test, the distribution of the product test, and the MCMC test further present the significance of mediation effect of moral hazard.

Robustness tests are conducted for large-scale farms. The mediation effect of moral hazards is still found on large-scale farms using the three-step method and the adjusted product of coefficients method. However, the mediation effect of moral hazard in large-scale farms using combine harvesting becomes insignificant due to the income incentive from the higher service fees of combine harvesters. After replacing the rice advantageous regional controls with rice cropping regional controls, the results are also similar to the above.

Therefore, this chapter illuminates the mediation path that the negative effect of harvest outsourcing services on operators' work attitudes will lead to increased rice harvest losses using mediation analysis models.

References

Angrist JD, Pischke J-S (2008) Mostly harmless econometrics: an empiricist's companion. Princeton University Press, Princeton

Aulakh J, Regmi A (2013) Post-harvest food losses estimation-development of consistent methodology. In: selected poster prepared for presentation at the agricultural and applied economics association's 2013 AAEA & CAES joint annual meeting. Washington DC

Bala BK, Haque MA, Hossain MA, Majumdar S (2010) Post harvest loss and technical efficiency of rice, wheat and maize production system: assessment and measures for strengthening food security. Bangladesh Agricultural University, Bengaluru, India

Baloch UK (1999) Wheat: post-harvest operations. Pakistan Agricultural Research Council, Islamabad

Baron RM, Kenny DA (1986) The moderator-mediator variable distinction in social psychological research: conceptual, strategic, and statistical considerations, pp 1173–1182

Basavaraja H, Mahajanashetti SB, Udagatti NC (2007) Economic analysis of post-harvest losses in food grains in India: a case study of Karnataka. Agric Econ Res Rev 20:117–126. https://doi.org/10.22004/ag.econ.47429

Begum EA, Hossain MI, Papanagiotou E (2012) Economic analysis of post-harvest losses in food grains for strengthening food security in northern regions of Bangladesh. Int J Appl Res Bus Adm Econ 01:56–65

Cai J, Liu WY (2019) Agricultural social service and opportunistic behavior: take agricultural machinery operation services as example. Reform 18–29

Cao FF, Huang D, Zhu JF, Wu LP (2018) The wheat harvest loss and its main determinants in China: an empirical analysis based on survey data from 1135 households. China Rural Surv 75–87

Chegere MJ (2017) Post-harvest losses, intimate partner violence and food security in Tanzania. University of Gothenburg, Gothenburg, Sweden

Chuan-udom S, Chinsuwan W (2010) Operating factors affecting harvesting losses of cleaning unit of rice combine harvesters. Asia-Pacific J Sci Technol 15:487–495

Danbaba N, Idakwo PY, Kassum AL et al (2019) Rice postharvest technology in Nigeria: an overview of vurrent status, constraints and potentials for sustainable development. Open Access Libr J 6:1–23. https://doi.org/10.4236/oalib.1105509

Deng X, De XuD, Zeng M, Bin QY (2020) Does outsourcing affect agricultural productivity of farmer households? Evidence from China. China Agric Econ Rev 12:673–688. https://doi.org/10.1108/CAER-12-2018-0236

Dercon S, Gilligan DO, Hoddinott J, Woldehanna T (2009) The impact of roads and agricultural extension on consumption growth and poverty in fifteen Ethiopian villages. Am J Agric Econ 91:1007–1021. https://doi.org/10.1111/j.1467-8276.2009.01325.x

Diener E, Suh EM, Lucas RE, Smith HL (1999) Subjective wellbeing: three decades of progress. Psychol Bull 125:276–302. https://doi.org/10.1037/0033-2909.125.2.276

Ei-Awady MN, Yehia I, Ebaid MT, Arif EM (2009) Development of rice cleaner for reduced impurities and losses. Agric Mech Asia, Afr Lat Am 40:15–20

Fan S, Hazell P, Thorat S (2000) Government spending, growth and poverty in rural India. Am J Agric Econ 82:1038–1051. https://doi.org/10.1111/0002-9092.00101

Feinberg FM (2012) Mediation analysis and categorical variables: some further frontiers. J Consum Psychol 22:595–598. https://doi.org/10.1016/j.jcps.2012.03.007

Fenn HRW, Laycock E (2017) A socio-economic investigation of pre-harvest and post-harvest crop loss between producers and retailers in Fenland. Sheff Hallam Univ Nat Environ Res Trans 3:38–55

Gao LW, Xu SW, Li ZM et al (2016) Main grain crops postharvest losses and its reducing potential in China. Trans Chinese Soc Agric Eng 32:1–11

Goldsmith PD, Martins AG, de Moura AD (2015) The economics of post-harvest loss: a case study of the new large soybean—maize producers in tropical Brazil. Food Secur 7:875–888. https://doi.org/10.1007/s12571-015-0483-4

Greeley M (1982) Farm-level post-harvest food losses: the myth of the soft third option. IDS Bull 13:51–60. https://doi.org/10.1111/j.1759-5436.1982.mp13003007.x

Han QL (2019) On the cohesion dilemma between family management and agricultural social service: based on M county. J Nanjing Agric Univ (Soc Sci Ed) 19:20–27

Hasan MK, Tanaka TST, Alam MM et al (2020) Impact of modern rice harvesting practices over traditional ones. Rev Agric Sci 8:89–108. https://doi.org/10.7831/ras.8.0_89

Hayes AF, Preacher KJ (2010) Quantifying and testing indirect effects in simple mediation models when the constituent paths are nonlinear. Multivariate Behav Res 45:627–660. https://doi.org/10.1080/00273171.2010.498290

Huan ML, Hou YX (2020) Quality control contract model of service in agricultural production outsourcing. J Agro-Forestry Econ Manag 19:288–296. https://doi.org/10.16195/j.cnki.cn36-1328/f.2020.03.31

Huang D, Yao L, Wu LP, Zhu X Di (2018) Measuring rice loss during harvest in China: based on experiment and survey in five provinces. J Nat Resour 33:1427–1438. https://doi.org/10.31497/zrzyxb.20170810

Iacobucci D (2012) Mediation analysis and categorical variables: the final frontier. J Consum Psychol 22:582–594. https://doi.org/10.1016/j.jcps.2012.03.006

Ji C, Guo HD, Jin SQ, Yang J (2017) Outsourcing agricultural production: evidence from rice farmers in Zhejiang province. PLoS ONE 12:1–16. https://doi.org/10.1371/journal.pone.0170861

Kantor LS, Lipton K, Manchester A, Oliveira V (1997) Estimating and addressing America's food losses. Food Rev Food Rev 20:2–12. https://doi.org/10.22004/ag.econ.234453

Kleibergen F, Paap R (2006) Generalized reduced rank tests using the singular value decomposition. J Econom 133:97–126. https://doi.org/10.1016/J.JECONOM.2005.02.011

Li HY, Zhao JH (2018) Rebound effects of new irrigation technologies: the role of water rights. Am J Agric Econ 100:786–808. https://doi.org/10.1093/ajae/aay001

Li ZF, Xia PK, Wang ZH et al (1991) Analysis of the constitution of grain postproduction losses and the preventive measures. J Zhejiang Univ 17:389–395

Li Y, Schneider julie A, Bennett david A (2007) Estimation of the mediation effect with a binary mediator. Stat Med 26:3398–3414. https://doi.org/10.1002/sim.2730.https://doi.org/10.1002/sim.2730

Li XF, Huang D, Qu X, Zhu JF (2020) Effects of different harvesting ways on grain loss: based on the field survey of 3251 rural households in China. J Nat Resour 35:1043–1054. https://doi.org/10.31497/zrzyxb.20200503

Lu QA, Du XD (2020) The outsourcing choice of agricultural production tasks: implications for food security—a multiple-task based approach. In: The 2020 agricultural and applied economics association annual meeting. Kansas City, Missouri

Mackinnon DP (2008) Introduction to statistical mediation analysis. Lawrence Erlbaum Associates, New York

MacKinnon DP, Fairchild AJ (2009) Current directions in mediation analysis. Curr Dir Psychol Sci 18:16–20. https://doi.org/10.1111/j.1467-8721.2009.01598.x

MacKinnon DP, Cox MG (2012) Commentary on "mediation analysis and categorical variables: the final frontier" by Dawn Iacobucci. J Consum Psychol 22:600–602. https://doi.org/10.1016/j.jcps.2012.03.009
MacKinnon DP, Lockwood CM, Williams J (2004) Confidence limits for the indirect effect: distribution of the product and resampling methods. Multivariate Behav Res 39:99. https://doi.org/10.1207/s15327906mbr3901_4
Martins AG, Goldsmith P, Moura A (2014) Managerial factors affecting post-harvest loss: the case of Mato Grosso Brazil. Int J Agric Manag 3:200–209. https://doi.org/10.5836/ijam/2014-04-03
Minor T, Astill G, Skorbiansky SR, et al (2020) Economic drivers of food loss at the farm and pre-retail sectors: a look at the produce supply chain in the United States. United States Department of Agriculture
Pandey V, Shanoyan A, Peterson HC, Ross RB (2013) Principal-agent governance mechanism in an emerging biofuels supply chain in USA. Asian J Agric Rural Dev 3:532–542
Parfitt J, Barthel M, MacNaughton S (2010) Food waste within food supply chains: quantification and potential for change to 2050. Philos Trans R Soc B Biol Sci 365:3065–3081. https://doi.org/10.1098/rstb.2010.0126
Picazo-Tadeo AJ, Reig-Martínez E (2006) Outsourcing and efficiency: the case of Spanish citrus farming. Agric Econ 35:213–222. https://doi.org/10.1111/j.1574-0862.2006.00154.x
Poungchompu S, Chantanop S (2016) Economic aspects of rice combine harvesting service for farmer in Northeast Thailand. Asian Soc Sci 12:201–211. https://doi.org/10.5539/ass.v12n8p201
Preacher KJ, Hayes AF (2004) SPSS and SAS procedures for estimating indirect effects in simple mediation models. Behav Res Meth Instrum Comput 36:717–731. https://doi.org/10.3758/BF03206553
Qu X, Kojima D, Nishihara Y et al (2020) Impact of rice harvest loss by mechanization or outsourcing: comparison of specialized and part-time farmers. Agric Econ (Czech Republic) 66:542–549. https://doi.org/10.17221/253/2020-AGRICECON
Qu X, Kojima D, Nishihara Y, et al (2021) A study of rice harvest losses in China : do mechanization and farming scale matter? Jpn J Agric Econ 23:83–88. https://doi.org/10.18480/jjae.23.0_83
ReFED (2016) A roadmap to reduce U.S. food waste by 20%. Rethink Food Waste Through Economics and Data, New York, USA
Segrè A, Falasconi L, Politano A, Vitturari M (2014) Background paper on the economics of food loss and waste. Food Agriculture Organization of United Nations, Rome, Italy
Shen HF, Chen C, Liao XY, Wang L (2015) Spatial econometric analysis of rice production links outsourcing pricing mechanism: empirical study of 14 province 42 counties in China. China Rural Surv 34–46
Smale M, Singh J, Di Falco S, Zambrano P (2008) Wheat breeding, productivity and slow variety change: evidence from the Punjab of India after the green revolution. Aust J Agric Resour Econ 52:419–432. https://doi.org/10.1111/j.1467-8489.2008.00435.x
Sun DQ, Rickaille M, Xu ZG (2018) Determinants and impacts of outsourcing pest and disease management: evidence from China's rice production. China Agric Econ Rev 10:443–461. https://doi.org/10.1108/CAER-01-2017-0011
Tobin J (1985) Estimation of relationships for limited dependent variables. Econom J Econom Soc 26:24–36. https://doi.org/10.2307/1907382
Wang GM, Yi ZY, Chen C, Cao GQ (2016) Effect of harvesting date on loss component characteristics of rice mechanical harvested in rice and wheat rotation area. Trans Chinese Soc Agric Eng 32:36–42
Wang KR, Xie RZ, Min B et al (2021) Review of combine harvester losses for maize and influencing factors. Int J Agric Biol Eng 14:1–10. https://doi.org/10.25165/j.ijabe.20211401.6034
Widawsky D, Rozelle S (1998) Varietal diversity and yield variability in Chinese rice production. Farmers gene banks and crop breeding: economic analyses of diversity in wheat maize and rice. Springer, Dordrecht, pp 159–172

Wu LH, Hu QP, Wang JH, Zhu D (2017) Empirical analysis of the main factors influencing rice harvest losses based on sampling survey data of ten provinces in China. China Agric Econ Rev 9:287–302. https://doi.org/10.1108/CAER-03-2016-0036

Wuepper D, Yesigat Ayenew H, Sauer J (2018) Social capital, income diversification and climate change adaptation: panel data evidence from rural Ethiopia. J Agric Econ 69:458–475. https://doi.org/10.1111/1477-9552.12237

Yang J, Huang ZH, Zhang XB, Reardon T (2013) The rapid rise of cross-regional agricultural mechanization services in China. Am J Agric Econ 95:1245–1251. https://doi.org/10.1093/ajae/aat027

Zhan YR (1995) Sampling survey and analysis of national grain post-harvest losses. Chian Grain Econ 4:44–47

Zhang XB, Yang J, Thomas R (2017) Mechanization outsourcing clusters and division of labor in Chinese agriculture. China Econ Rev 43:184–195. https://doi.org/10.1016/j.chieco.2017.01.012

Chapter 6
The Effect of Moral Hazard on Rice Harvest Loss

In Chap. 5, the mediation path that the negative effect of harvest outsourcing services on operators' work attitudes will lead to increased rice harvest losses is examined through mediation analysis models. In the previous chapters, the effect of service providers' moral hazards on rice harvest losses has not been studied directly. In this chapter, we focus only on farmers using harvest outsourcing services and directly explore the effect of service providers' moral hazards on rice harvest losses.

6.1 Introduction

In China, organizations or individuals are engaged in providing agricultural machinery outsourcing services to farmers who do not own machines. This enables small-scale farms to use machinery for agricultural production without owning it (Ji et al. 2017). Agricultural machinery outsourcing service is considered a way to achieve economies of scale (Cai and Wang 2021). It has solved the mechanization challenges faced by small-scale farms, enabling mechanization of certain crop production stages, such as harvesting, to be conducted on a much larger scale, although the scale of production is small at the farm level (Picazo-Tadeo and Reig-Martínez 2006; Zhang et al. 2017). Similar practices are observed in various countries characterized by small-scale farms. In the Netherlands, small-scale farmers lacking sufficient labor prefer to buy agricultural outsourcing services (Igata et al. 2008).

The booming harvest outsourcing services have received considerable attention, the origin and development of agricultural machinery outsourcing services (Yang et al. 2013), farmers' outsourcing service participation and its influencing factors (Cai and Wang 2021; Yi 2018), and the effect of outsourcing services on agricultural production (Deng et al. 2020; Lu and Du 2020) and farmers' welfare (Mi et al. 2020). A few studies have noticed the changes in the relationship between farmers and market caused by harvest outsourcing services. The essence of harvest outsourcing

X. Qu et al., *Harvest Loss in China*, The University of Tokyo Studies on Asia,
https://doi.org/10.1007/978-981-97-9156-9_6

services is the division of labor (Yi 2018). The relationship between farmers and service providers is that of principal and agent. Some scholars have elaborated on the possibility of service providers' moral hazards from the perspective of game theory (Cai and Liu 2019; Huan and Hou 2020). Huan and Hou (2020) argued that the goal of service providers is to maximize their own profits without considering factors such as crop yield and quality, land fertility decline, and pollution. Therefore, careless practices by service providers will harm farmers' interests. One of the direct consequences of service provider' moral hazards is an increase in harvest losses. Service providers' moral hazard practices, such as increasing forward speed of machines or leaving some crops in the corner unharvested, directly contribute to increased harvest losses. However, there is currently a lack of empirical evidence to determine whether service providers' moral hazards will have a substantial effect on agricultural production. Therefore, this chapter will directly investigate the effect of service providers' moral hazards on rice harvest losses.

As shown in Chap. 4, service providers' work attitudes become more serious when serving large-scale farms and business farms. Based on this, we assume that the effect of service providers' moral hazards on rice harvest losses may vary with farming scale and part-time farming level. Therefore, similar to Chap. 4, the chapter will examine service providers' moral hazards on rice harvest losses from farming scale and part-time farming perspectives. Given the critical role of large-scale farms and business farms in rice production, this chapter will focus particularly on them.

6.2 Data and Method

6.2.1 *Data and Variable*

The data used in this chapter is also derived from the surveys introduced in Chap. 3.[1] Unlike Chaps. 4 and 5, this chapter examines 651 households that used harvest outsourcing services. The sample still covers most of the three rice advantageous regions (Table 6.1). For provinces that do not belong to any of these three advantageous regions, they are classified into corresponding advantageous regions according to their natural resources, cultivation characteristics, and geographical location. Compared to Table 3.1, only Heilongjiang province in the Northeast Plain and Guangdong province in the Southeast Coast are not included.

6.2.1.1 Dependent Variable

As described in Chap. 3, the starting and ending points of harvest losses in this study are defined as the field and storage places, respectively. The stages during which

[1] Please check Chap. 3 for the detail information about the dataset.

Table 6.1 Sample distribution based on advantageous regions

Regions	Sample provinces	Sample
The Northeast Plain	Tianjin, **Liaoning, Jilin**, and Shandong	52 (7.99%)
The Yangtze River Basin	**Jiangsu, Anhui, Jiangxi, Hubei, Hunan, Sichuan, Guizhou, Yunnan, Chongqing**, and Shaanxi	492 (75.58%)
The Southeast Coast	**Zhejiang, Fujian**, and **Guangxi**	107 (16.43%)

Note The provinces that are bolded are those belong to the corresponding advantageous areas based on the *Planning*
Data source Planning and author's calculation based on the survey

losses occur are reaping, threshing, winnowing, and field transport. To facilitate comparison among farms of different scales, harvest loss rate is used to measure harvest losses.

$$\text{HLR} = \frac{\text{harvest losses}}{\text{harvest losses} + \text{PRO}} = \frac{L_{\text{reap}} + L_{\text{thr}} + L_{\text{win}} + L_{\text{tra}}}{\left(L_{\text{reap}} + L_{\text{thr}} + L_{\text{win}} + L_{\text{tra}}\right) + \text{PRO}} \times 100\% \tag{6.1}$$

As noted in Chap. 5, if the harvest loss per unit area was less than the value at the 10th percentile, it is replaced by the value at the 10th percentile. We believe that such a treatment is reasonable because it is impossible to completely avoid harvest losses with current technology.

6.2.1.2 Core Independent Variable

In Chap. 4, operators' work attitudes are used as the proxy for the level of effort. If the service provider's work attitude is less serious than that of farmer, it is considered that there is a reduced effort level. Slightly different from the approach in Chap. 4, work attitudes are directly used to measure the reduced effort level. As shown in Eq. (6.2),[2] if the service provider's work attitude was estimated to be serious, the service provider was considered to have no reduced effort level, i.e., no moral hazard and WA equals 1; if the service provider's work attitude was not serious, the service provider was considered to have reduced effort level, i.e., moral hazards and WA equals 0.

$$\text{WA} = \begin{cases} 0, & \text{if service provider's work attitude was not serious (moral hazard)} \\ 1, & \text{if service provider's work attitude was serious (no moral hazard)} \end{cases} \tag{6.2}$$

[2] Because we focus on the farms using harvest outsourcing services, the work attitude here only refers to service provider's work attitude. This is different with Chap. 4.

6.2.1.3 Covariates

Based on existing studies, we add the following three types of control variables: (1) production and harvesting conditions, (2) household and individual characteristics, and (3) regional control variables. These three types of control variables are the same as those in Chap. 5.

6.2.2 *Two-Stage Least Squares*

To investigate the relationship between rice harvest losses and service providers' work attitudes, we consider a multiple regression model with the following relationship:

$$\mathrm{HLR} = \alpha_0 + \alpha_1(WA) + \sum_{i=2}^{k} \alpha_i c_i + u, \tag{6.3}$$

where HLR is rice harvest loss rates. WA represents service providers' work attitudes. $c_2, \ldots c_i$ are control variables that have effects on rice harvest loss rates, which are listed in the "Covariates". $\alpha_0, \ldots \alpha_i$ are the parameters to be estimated. u is the error term, which is a mean zero random variable that is independent of explanatory variables.

One concern in the estimation through this model is reverse causality. Work attitudes may be endogenous. High harvest losses may increase the probability of moral hazards. On rugged or irregularly shaped plots, harvest losses are usually high, which provides conditions for service providers to inadvertently create moral hazards or to their inherent moral hazards. Because it is difficult to distinguish whether the high harvest losses are caused by the plot's terrain or by moral hazards. This creates conditions for service providers to implement moral hazard practices. Moreover, since estimations of operators' work attitudes were collected after the harvesting operation is completed, farmers might estimate operators' work attitude based on harvest losses. Such mutual causation leads to endogeneity problem, violating the independent assumption of the error term; the explanatory variables would be correlated with the error term, resulting in biased estimation outcomes.

To address the endogeneity issue of work attitudes, the instrumental variable used in Chap. 5 is introduced: the distance from homestead to the nearest paved road (*Htor*). As indicated in the following two points, this instrumental variable is related to work attitudes. Firstly, the further the distance from homestead to the nearest paved road, the longer the travel distance on unpaved road, which requires careful driving. Secondly, farmers who live in areas far from paved roads are more likely to be elder farmers and economically disadvantaged farmers (Dercon et al. 2009; Fan et al. 2000). Their aversion to food loss (Greeley 1986) can affect service providers' work attitudes toward harvesting work. Moreover, the exclusion restriction seems to be met. The homestead locations and paved roads planned by the village communities determine the instrumental variable, making it unlikely to affect harvest losses in the fields. Therefore, we consider the distance from homestead to the nearest paved road as a credible instrumental variable. Then, the model is estimated in an instrumental variable framework:

$$\text{HLR} = \alpha_0 + \alpha_1\left(\widehat{\text{WA}}\right) + \sum_{i=2}^{k} \alpha_i c_i + u, \tag{6.4}$$

$$\text{WA} = \beta_0 + \beta_1(\text{Htor}) + \sum_{i=2}^{k} \beta_i c_{1i} + v, \tag{6.5}$$

where Htor is the instrumental variable that is correlated with work attitude but influences rice harvest loss rates only through work attitudes. $c_{12}, \ldots c_{1i}$ represent control variables that have effects on operators' work attitudes, which are the same as those in Chap. 4. v is the error term, which is independent of explanatory variables. Then 2SLS is used for the estimation.

6.2.3 Classification of Farm Type

6.2.3.1 Large-Scale Farms

In this chapter, farms are divided in the same way as used in Chap. 4. Using the median of rice faming scale (0.267 ha), farms are divided into small-scale farms and large-scale farms. As given in Table 6.2, farms with a rice planting area of less than 0.267 ha are classified as small-scale farms, while farms with a rice planting area greater than or equal to 0.267 ha are classified as large-scale farms. Furthermore, we divide the sample into ten groups according to the deciles of the rice planting area and conduct the analysis on the largest decile of farms.

Table 6.2 Definitions of different farming scales

Classification	Definition
Small-scale farm	Rice planting area < the median of the rice planting area (0.267 ha)
Large-scale farm	Rice planting area ≥ the median of the rice planting area (0.267 ha)
Largest decile of farm	Rice planting area > the ninth decile of the rice planting area

Notes (1) Samples that used manual reaping and manual threshing were not included when counting large-scale farms, as large-scale farms are unlikely to adopt these methods
Data source Author's calculation based on the survey

Table 6.3 Definitions of business farms and rice-dominated farms

Classification	Definition
Business farm	Farm income/total income > 1/2
Rice-dominated farm	Farm income/total income > 1/2 and biggest source of income is rice

Note (1) In this study, the proportion of rural household management income is used to represent the proportion of farm income
Data source Author's calculation based on the survey

6.2.3.2 Business Farms

Similar to Chap. 4, business farms here refer to farms where income from farming accounts for more than 50% of their total income. Therefore, business farms will invest more resource and time in farming compared to in non-farming.

To diversify risks and meet diversified household consumption needs, farmers often choose to grow a variety of agricultural production. With limited resources and time, they allocate more production resources to the crop that accounts for the largest share of farm income. If business farms would devote more resources and time to agricultural production, then business farms with rice as their highest source of income would be more concerned with rice production. Therefore, we further conduct the analysis on rice-dominated farms, where biggest source of income is rice. Table 6.3 gives the definitions of business farms and rice-dominated farms.

6.3 Results and Discussion

6.3.1 651 Farms

6.3.1.1 Variable Description Statistics

Table 6.4 gives the summary and definitions of variables for farms that purchased harvest outsourcing services. Overall, the average rice harvest loss rate for farmers who purchased harvest outsourcing services was 3.54%. Only 17% of service providers had a serious attitude. The adoption rates of combine harvesters, mechanical winnowing, and mechanical field transport were 76%, 53%, and 71%, respectively. The average rice planting area was 0.40 ha, which was larger than that for 1106 full sample (0.33 ha in Table 4.2). It means that farmers who purchased outsourcing services had a relatively large planting area.

Table 6.4 Summary and definition of variables (651 farms)

Variable		651 farms	Std. Dev
Dependent variable			
HLR	Harvest loss rate (%)	3.54	3.37
Core independent variable			
WA	Dummy = 1 if service provider's harvest attitude was serious, 0 otherwise	0.17	0.38
Instrumental variable			
Htor	Distance from homestead to the nearest paved road	0.30	0.68
Production and harvesting variable			
Com	Dummy = 1 if combine harvesting, 0 if segmented harvesting	0.76	0.43
Win	Dummy = 1 if machinery winnowing, 0 otherwise	0.53	0.50
Tra	Dummy = 1 if machinery field transport, 0 otherwise	0.71	0.45
Wea	Dummy = 1 if bad weather when harvesting, 0 if normal weather	0.14	0.35
Pest	No pest = 1, slight pests = 2, general or serious pests = 3	1.79	0.78
Area	Rice planting area (ha)	0.40	0.40
Yield	Yield (quintal/ ha)	84.81	26.36
Flat	Dummy = 1 if plot terrain is flat, 0 otherwise	0.81	0.40
Dis	Distance from the field to storage locations (km)	0.58	0.62
Labor	Dummy = 1 if farmer reported a lack of manpower, 0 otherwise	0.23	0.42
Sav	Dummy = 1 if farmer picked up rice left in field, 0 otherwise	0.18	0.38
Mat	Dummy = 1 if rice was mature when harvesting, 0 otherwise	0.93	0.25
Price	Sale price of rice (CNY/kg)	2.88	0.21
Household and individual variable			
Gen	Gender of the household head (male = 1, female = 0)	0.87	0.33
Age	Age of household head	54.33	10.06
Edu	School years of household head (years)	7.33	2.59
Train	Dummy = 1 if household head had agricultural training, 0 otherwise	0.10	0.29
Tinc	Household income (ten thousand CNY)	7.21	5.67
Rincs	Rice income as a percentage of total income (%)	19.42	20.41
N		651	

Note (1) The highest correlation coefficient between independent variables is no more than 0.8 and the VIF-value is less than 10, which means there is no multicollinearity
Data source Author's calculation based on the survey

6.3.1.2 Estimation Results

If the endogeneity of work attitude does not exist, there is no need to estimate the 2SLS regression because the OLS regression would provide consistent and more valid estimates (Kaufmann et al. 2019). Table 6.5 gives the tests for the instrumental variable. First, the p-value for the endogeneity test given by the "ivreg2" command in Stata is 0.022, indicating the necessity to address the endogeneity of work attitude. Then, the coefficient of the instrumental variable (Htor) in the first stage of 2SLS is 0.145, which is significant at the 1% significance level. Meanwhile, the Kleibergen–Paap rk Wald F statistic (Kleibergen and Paap 2006) is 25.311, higher than the 10% critical value (16.38). Therefore, the weakness of the instrumental variable is rejected.

Moreover, Wuepper et al. (2018) indicated that the more the instrumental variable correlates with other control variables, the higher the possibility of failing the exclusion restriction. Therefore, we test whether the instrumental variable correlates with plot terrain (Flat) and rice planting area (Area). The estimation results in columns (2) and (3) of Table 6.5 show that the instrumental variable is not significantly correlated with them. Therefore, the estimation results of 2SLS are more reliable than those of OLS.

Table 6.6 displays the estimation results of OLS and 2SLS regressions for 651 farms using outsourcing services. Column (2) shows the estimation results of 2SLS

Table 6.5 Tests of instrumental variable

	Work attitude	(2) Flat	(3) Area
	(1) First stage (2SLS)		
Htor	0.145^{***} (0.03)	0.279 (0.21)	0.003 (0.02)
Constant	0.592^{**} (0.29)	3.662^{***} (0.99)	0.394^{***} (0.11)
Production and production control	Yes	No	No
Household and individual control	Yes	Yes	Yes
Advantageous region	Yes	Yes	Yes
Kleibergen–Paap F statistic	25.311		
Endogeneity test	$p = 0.022$		
Observations	651	651	651

Notes (1) ***, **, and * indicate 1, 5, and 10% significance levels, respectively. (2) Robust standard errors statistics are in parentheses. (3) Stock–Yogo critical values for weak identification tests (used for Kleibergen–Paap rk Wald F statistics) are 16.38 for 10% maximal IV bias; 8.96 for 15% maximal IV bias; 6.66 for 20% maximal IV bias; and 5.53 for 25% maximal IV bias (Stock and Yogo 2005). (4) The results of other variables can be found in column (1) of Table A6 in Appendices. 5) Logit regression and ordinary least squares are used for estimation of column (2) and (3), respectively
Data source Author's calculation based on the survey

regression. The coefficient of work attitude (WA) is negative—3.082, which is significant at the 1% significance level. This implies that service providers' serious work attitudes will decrease rice harvest loss rates. In other word, service providers' moral hazards will increase rice harvest loss rates. For comparison, column (1) lists the estimation results of OLS regression. Without dealing with the endogeneity of work attitudes, the coefficient for work attitude (WA) estimated by OLS regression is –0.629, which is significant at the 10% significance level. Therefore, both the absolute value and significance level given by 2SLS regression are higher than those by OLS regression; this indicates that failing to account for the endogeneity of work attitudes leads to an underestimation of the true effect of moral hazards on rice harvest loss rates.

Moreover, mechanical winnowing (Win), bad weather (Wea), pests (Pest), flat terrain (Flat), shortage of labor (Labor), food saving consciousness (Sav), and training (Tra) have significantly positive coefficients on harvest loss rates, while the yield (Yield), sale price (Price), and the percentage of rice income (Rincs) have significantly negative coefficients.

6.3.2 *Farming Scale Perspective*

6.3.2.1 Variable Description Statistics

Table 6.7 gives the summary and definitions of variables for farms of different scales. On small-scale farms, the average rice harvest loss rate was 4.33%, which was higher than that on large-scale farms (2.72%). On the largest decile of farms in terms of area, the average rice harvest loss rate was further reduced (2.06%). The possible reason is that larger areas facilitate mechanical operations and access, thereby reducing harvest losses. Only 11% of service providers took the harvesting seriously on small-scale farms, compared to 24% and 31% of service providers had serious work attitudes on large-scale farms and the largest decile of farms, respectively. The average rice planting area for large-scale farms was 0.66 ha, which was more than four times that for small-scale farms and less than half of that for the largest decile of farms.

The average age of household heads of the large-scale farms was 52.76 years old, which was 3.1 years younger than that of small-scale farms and 1.11 years older than that of the largest decile of farms, respectively. There was little difference in schooling years between these three groups of household heads. The household heads of the largest decile of farms had the shortest average years of education (6.91 years), which was 0.47 years shorter than that of large-scale farmers and 0.37 years shorter than that of small-scale farmers, respectively. This is only equivalent to junior high school education. With the expansion of rice planting area, both household income and the share of rice income increased. In particular, the share of rice income was 50.82% on average for the largest decile of farms, which was approximately 5.47 times that of small-scale farms and about 1.71 times that of large-scale farms.

Table 6.6 Estimation results of work attitude on HLR (651 farms)

	Rice harvest loss rate (%)			
	(1) OLS		(2) 2SLS	
Core independent variable				
WA	−0.629*	(0.33)	−3.082***	(1.07)
Production and harvesting variable				
Com	0.236	(0.33)	0.567	(0.37)
Win	1.253***	(0.27)	1.302***	(0.27)
Tra	−0.550*	(0.30)	−0.423	(0.30)
Wea	0.684	(0.49)	0.867*	(0.52)
Pest = 2	0.544**	(0.27)	0.398	(0.29)
Pest = 3	1.201***	(0.37)	0.862**	(0.40)
Area	−1.119***	(0.38)	−0.665	(0.43)
Yield	−0.014**	(0.01)	−0.015***	(0.01)
Flat	0.772**	(0.35)	0.791**	(0.36)
Dis	−0.238	(0.19)	−0.212	(0.19)
Labor	0.862***	(0.29)	0.610*	(0.32)
Sav	1.125***	(0.33)	1.200***	(0.33)
Mat	0.030	(0.45)	0.189	(0.48)
Price	−1.468**	(0.61)	−1.756***	(0.63)
Household and individual variable				
Gen	0.614**	(0.31)	0.521	(0.33)
Age	0.018	(0.01)	0.017	(0.01)
Edu	−0.051	(0.06)	−0.054	(0.06)
Train	1.209**	(0.49)	1.197**	(0.51)
Tinc	0.025	(0.03)	0.018	(0.03)
Rincs	−0.016**	(0.01)	−0.018**	(0.01)
Cons	8.170***	(2.17)	8.694***	(2.23)
Advantageous region	Yes		Yes	
R^2	0.23			
N	651		651	

Notes (1) ***, **, and * indicate 1, 5, and 10% significance levels, respectively. (2) Robust standard errors statistics are in parentheses
Data source Author's calculation based on the survey

Table 6.7 Summary and definition of variables (farming scale)

Variable		Small-scale farms	Large-scale farms	Largest decile of farms
Dependent variable				
HLR	Harvest loss rate (%)	4.33	2.72	2.06
Core independent variable				
WA	Dummy = 1 if service provider's harvest attitude was serious, 0 otherwise	0.11	0.24	0.31
Instrumental variable				
Htor	Distance from homestead to the nearest paved road	0.33	0.28	0.44
Production and harvesting variable				
Com	Dummy = 1 if combine harvesting, 0 if segmented harvesting	0.76	0.75	0.80
Win	Dummy = 1 if machinery winnowing, 0 otherwise	0.62	0.45	0.23
Tra	Dummy = 1 if machinery field transport, 0 otherwise	0.61	0.82	0.92
Wea	Dummy = 1 if bad weather when harvesting, 0 if normal weather	0.08	0.20	0.23
Pest	No pest = 1, slight pests = 2, general or serious pests = 3	1.70	1.89	1.74
Area	Rice planting area (ha)	0.15	0.66	1.39
Yield	Yield (quintal/ ha)	86.00	83.60	80.45
Flat	Dummy = 1 if plot terrain is flat, 0 otherwise	0.86	0.75	0.89
Dis	Distance from the field to storage locations (km)	0.54	0.63	0.77
Labor	Dummy = 1 if farmer reported a lack of manpower, 0 otherwise	0.28	0.18	0.25

(continued)

Table 6.7 (continued)

Variable		Small-scale farms	Large-scale farms	Largest decile of farms
Sav	Dummy = 1 if farmer picked up rice left in field, 0 otherwise	0.19	0.17	0.29
Mat	Dummy = 1 if rice was mature when harvesting, 0 otherwise	0.96	0.91	0.97
Price	Sale price of rice (CNY/kg)	2.90	2.86	2.90
Household and individual variable				
Gen	Gender of the household head (male = 1, female = 0)	0.86	0.89	0.94
Age	Age of household head	55.86	52.76	51.65
Edu	School years of household head (years)	7.28	7.38	6.91
Train	Dummy = 1 if household head had agricultural training, 0 otherwise	0.12	0.07	0.15
Tinc	Household income (ten thousand CNY)	6.57	7.87	8.08
Rincs	Rice income as a percentage of total income (%)	9.29	29.77	50.82
N		329	322	65

Note (1) The highest correlation coefficient between independent variables is no more than 0.8 and the VIF-value is less than 10, which means there is no multicollinearity
Data source Author's calculation based on the survey

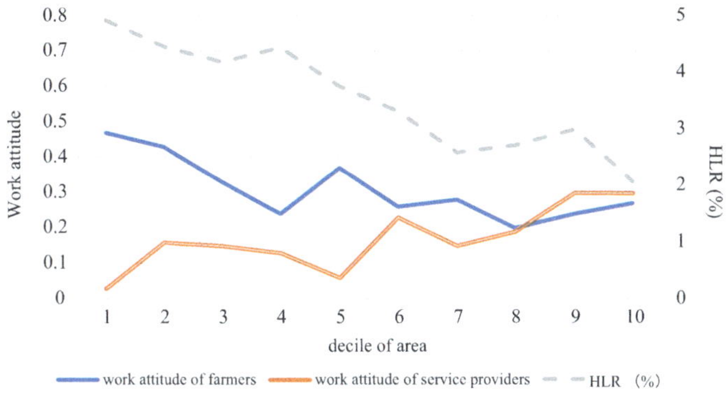

Fig. 6.1 Average work attitudes of service providers in different farm scales

To further understand the variation in service providers' work attitudes and rice harvest loss rates across different farm scales, we divide the sample into 10 groups according to the deciles of rice planting area. As shown in Fig. 6.1, generally, rice harvest loss rates decreased with the increase in farm scale. For service providers, their service fees are proportional to the serviced area, and in many cases, they face increasing return to scale. Thus, as farm scale increases, service providers might become more diligent due to the incentives of increased income. Farming management becomes difficult as farm scale increases, resulting in less serious work attitudes among farmers. When farm scale was small, farmers on average had a more serious work attitude than service providers. As farm scale expanded, the gap in average work attitudes between farmers and service providers gradually became narrow.

6.3.2.2 Estimation Results

Table 6.8 presents the estimation results of OLS and 2SLS regressions for small-scale farms and large-scale farms. The *p*-values for the endogeneity tests are 0.002 and 0.045 for small-scale farms and large-scale farms, respectively. Therefore, the 2SLS regression would have better results than OLS regression. The Kleibergen–Paap rk Wald *F* statistics (Kleibergen and Paap 2006) is above the 15% critical value (8.96) for small-scale farms (11.745) and above the 10% critical value (16.38) for large-scale farms (23.467). Overall, the weakness of the instrumental variable is rejected, although the evidence is not overwhelming.

Columns (2) and (4) present the estimation results of 2SLS regressions. After dealing with the endogeneity, the coefficient for work attitude (WA) is negative (−7.968) and significant at the 1% significance level on small-scale farms. Meanwhile, on large-scale farms, the coefficient for work attitude (WA) is negative (−1.746) and significant at the 10% significance level. This means that a serious work attitude will decrease rice harvest loss rate for both small-scale and large-scale farms. Moreover, the effect of work attitude on rice harvest loss rate is greater for

Table 6.8 Estimation results of work attitude on HLR (small- and large-scale farms)

Variable	Rice harvest loss rate (%)							
	Small-scale farms				Large-scale farms			
	(1) OLS		(2) 2SLS		(3) OLS		(4) 2SLS	
Core independent variable								
WA	−1.494***	(0.45)	−7.968***	(2.51)	0.159	(0.44)	−1.746*	(0.93)
Production and harvesting variable								
Com	0.815*	(0.49)	1.147**	(0.55)	−0.487	(0.43)	−0.109	(0.48)
Win	1.662***	(0.38)	1.270***	(0.44)	0.863**	(0.35)	1.045***	(0.36)
Tra	−0.507	(0.40)	−0.331	(0.44)	−0.049	(0.44)	0.027	(0.45)
Wea	0.793	(0.74)	0.731	(0.77)	0.903	(0.66)	1.054	(0.67)
Pest = 2	0.588	(0.41)	0.182	(0.49)	0.637*	(0.35)	0.487	(0.36)
Pest = 3	1.738***	(0.55)	1.251*	(0.64)	1.045**	(0.51)	0.589	(0.55)
Area	−8.876***	(2.93)	−8.099**	(3.28)	−0.614*	(0.35)	−0.324	(0.36)
Yield	−0.027***	(0.01)	−0.023**	(0.01)	−0.012*	(0.01)	−0.014*	(0.01)
Flat	−0.154	(0.61)	0.290	(0.65)	0.888**	(0.44)	0.871*	(0.45)
Dis	−0.087	(0.37)	0.231	(0.53)	0.033	(0.23)	0.001	(0.22)
Labor	1.352***	(0.41)	0.649	(0.55)	0.033	(0.32)	−0.126	(0.34)
Sav	1.358***	(0.49)	1.792***	(0.59)	0.310	(0.40)	0.282	(0.40)
Mat	−1.234	(0.77)	−1.018	(0.97)	−0.252	(0.54)	0.048	(0.55)
Price	−1.236	(0.89)	−2.093*	(1.08)	−2.416***	(0.89)	−2.362***	(0.92)
Household and individual variable								
Gen	0.912*	(0.47)	0.699	(0.67)	0.361	(0.34)	0.297	(0.33)
Age	0.030	(0.02)	0.031	(0.02)	0.004	(0.02)	0.003	(0.02)

(continued)

Table 6.8 (continued)

Variable	Rice harvest loss rate (%)							
	Small-scale farms				Large-scale farms			
	(1) OLS		(2) 2SLS		(3) OLS		(4) 2SLS	
Edu	0.021	(0.08)	−0.006	(0.09)	−0.107	(0.07)	−0.109	(0.08)
Train	1.286*	(0.72)	1.646**	(0.82)	−0.246	(0.50)	−0.305	(0.58)
Tinc	0.021	(0.05)	−0.007	(0.05)	0.047	(0.03)	0.037	(0.03)
Rincs	−0.002	(0.02)	−0.016	(0.03)	−0.022***	(0.01)	−0.024***	(0.01)
Cons	7.462**	(3.12)	9.348**	(3.71)	13.237***	(3.10)	12.811***	(3.18)
Advantageous region	Yes		Yes		Yes		Yes	
R^2	0.31				0.22			
F statistic			11.745				23.467	
Endogeneity test			$p = 0.002$				$p = 0.045$	
N	329		329		322		322	

Notes (1) ***, **, and * indicate 1, 5, and 10% significance levels, respectively. (2) Robust standard errors statistics are in parentheses. (3) Stock–Yogo critical values for weak identification tests (used for Kleibergen–Paap rk Wald *F* statistics) are 16.38 for 10% maximal IV bias; 8.96 for 15% maximal IV bias; 6.66 for 20% maximal IV bias; and 5.53 for 25% maximal IV bias (Stock and Yogo 2005). (4) The first stage results can be found in column (1) and (2) of Table A5 in Appendices

Data source Author's calculation based on the survey

small-scale farms than for large-scale farms. As service providers become more diligent in their work attitudes (As shown in Fig. 6.2), the effect of moral hazards diminishes.

Moreover, columns (1) and (3) present the estimation results of OLS regressions for comparison. It is evident that OLS results greatly underestimate the true effect of moral hazard.

Table 6.9 gives the estimation results of OLS and 2SLS regressions for the largest decile of farms. The *p*-value of endogeneity test is 0.029, implying the need to address the endogeneity issue of work attitude. The Kleibergen–Paap rk Wald *F* statistic (Kleibergen and Paap 2006) is 30.452, which is larger than the 10% Stock–Yogo critical value (16.38), suggesting that the instrumental variable is not weak. Therefore, the results of 2SLS regression are more reliable than those of OLS regression. The coefficient for work attitude (WA) given by 2SLS regression is negative (−1.391) and insignificant. Therefore, work attitude has no effect on rice harvest loss rates for the largest decile of farms.

In addition, the change in coefficients of farmland terrain also indicates that service providers' moral hazards gradually decrease with the expansion of farming scale. The coefficient for the terrain (Flat) changes from positive (0.871) in Table 6.8 to negative (−4.759) in Table 6.9. In large-scale farms, flat terrain increases the rice harvest loss rate may be due to the fact that it facilitates service providers to speed up machine. As moral hazard issues diminish in the largest decile of farms, flat terrain creates favorable conditions for mechanical harvesting, thereby helping to reduce rice harvest loss. In the largest decile of farms, moral hazard problem is no longer an important factor affecting harvest losses. It is replaced by factors such as means of transportation and terrain.

6.3.3 *Business Farming Perspective*

6.3.3.1 Variable Description Statistics

Table 6.10 gives the summary and definitions of variables for business farms and rice-dominated farms. The average rice harvest loss rate for rice-dominated farms was 2.46%, while that for business farms was 2.86%. The average work attitudes of service providers for business farms and rice-dominated farms were 0.20 and 0.21, respectively. The average planting area for rice-dominated farms was 0.64 ha, while that on business farms was 0.57 ha. In business farms, rice income accounted for an average of 30.47%, while it accounted for 35.41% in rice-dominated farms. These indicate that rice production was more important in rice-dominated farms.

Table 6.9 Estimation results of the largest decile of the farms

	Rice harvest loss rate (%)			
	(2) OLS		(2) 2SLS	
Core independent variable				
WA	0.155	(0.56)	−1.391	(0.89)
Production and harvesting variable				
Com	−1.076	(0.85)	−0.800	(0.77)
Win	3.189***	(0.87)	3.380***	(0.83)
Tra	4.209**	(1.80)	4.964***	(1.65)
Wea	0.763	(0.98)	0.529	(0.96)
Pest = 2	1.410***	(0.51)	1.013**	(0.51)
Pest = 3	2.359***	(0.79)	1.858**	(0.78)
Area	−0.672	(0.55)	−0.837	(0.54)
Yield	−0.002	(0.02)	0.006	(0.02)
Flat	−3.800***	(1.36)	−4.759***	(1.43)
Dis	0.090	(0.56)	0.192	(0.45)
Labor	−2.120**	(0.80)	−2.388***	(0.68)
Sav	−1.168**	(0.54)	−0.948**	(0.47)
Mat	3.222**	(1.52)	3.849***	(1.32)
Price	−0.574	(1.89)	−2.233	(1.89)
Household and individual variable				
Gen	0.913	(0.66)	0.923*	(0.52)
Age	−0.000	(0.04)	−0.007	(0.03)
Edu	0.218**	(0.11)	0.182**	(0.08)
Train	−2.450***	(0.81)	−2.146***	(0.74)
Tinc	−0.083	(0.06)	−0.144**	(0.07)
Rincs	−0.034**	(0.01)	−0.028**	(0.01)
Cons	2.135	(7.06)	6.806	(6.81)
Advantageous region	Yes		Yes	
R^2	0.76			
F statistic			30.452	
Endogeneity test			$p = 0.029$	
N	65		65	

Notes (1) ***, **, and * indicate 1, 5, and 10% significance levels, respectively. (2) Robust standard errors statistics are in parentheses. (3) Stock–Yogo critical values for weak identification tests (used for Kleibergen–Paap rk Wald *F* statistics) are 16.38 for 10% maximal IV bias; 8.96 for 15% maximal IV bias; 6.66 for 20% maximal IV bias; and 5.53 for 25% maximal IV bias (Stock and Yogo 2005). (4) The first stage results can be found in column (3) of Table A5 in Appendices
Data source Author's calculation based on the survey

Table 6.10 Summary and definition of variables (business farms and rice-dominated farms)

Variable		Business farms	Rice-dominated farms
Dependent variable			
HLR	Harvest loss rate (%)	2.86	2.46
Core independent variable			
WA	Dummy = 1 if service provider's work attitude was serious, 0 otherwise	0.20	0.21
Instrumental variable			
Htor	Distance from homestead to the nearest paved road	0.39	0.37
Production and harvesting variable			
Com	Dummy = 1 if combine harvesting, 0 if segmented harvesting	0.81	0.85
Win	Dummy = 1 if machinery winnowing, 0 otherwise	0.39	0.35
Tra	Dummy = 1 if machinery field transport, 0 otherwise	0.78	0.83
Wea	Dummy = 1 if bad weather when harvesting, 0 if normal weather	0.15	0.17
Pest	No pest = 1, slight pests = 2, general or serious pests = 3	1.79	1.79
Area	Rice planting area (ha)	0.57	0.64
Yield	Yield (quintal/ ha)	85.16	86.92
Flat	Dummy = 1 if plot terrain is flat, 0 otherwise	0.82	0.84
Dis	Distance from the field to storage locations (km)	0.61	0.61
Labor	Dummy = 1 if farmer reported a lack of manpower, 0 otherwise	0.19	0.18
Sav	Dummy = 1 if farmer picked up rice left in field, 0 otherwise	0.20	0.20
Mat	Dummy = 1 if rice was mature when harvesting, 0 otherwise	0.93	0.93
Price	Sale price of rice (CNY/kg)	2.87	2.87
Household and individual variable			
Gen	Gender of the household head (male = 1, female = 0)	0.90	0.91
Age	Age of household head	53.94	54.21
Edu	School years of household head (years)	7.28	7.12
Train	Dummy = 1 if household head had agricultural training, 0 otherwise	0.07	0.06
Tinc	Household income (ten thousand CNY)	7.40	7.39
Rincs	Rice income as a percentage of total income (%)	30.47	35.41
N		275	215

Note (1) The highest correlation coefficient between independent variables is no more than 0.8 and the VIF-value is less than 10, which means there is no multicollinearity
Data source Author's calculation based on the survey

6.3.3.2 Estimation Results

Table 6.11 gives the estimation results for business farms and rice-dominated farms. The p-values of the endogeneity test at the bottom of Table 6.11 are 0.043 for business farms and 0.065 for rice-dominated farms, indicating the presence of endogeneity of work attitudes. The Kleibergen–Paap rk Wald F statistic (Kleibergen and Paap 2006) for business farms and rice-dominated farms are 27.314 and 32.071, respectively. They are both above the 10% critical value (16.38), implying that the instrumental variable may not be regarded as weak. Therefore, the estimation results by 2SLS regressions are more reliable than those by OLS regressions. After controlling for the endogeneity of work attitude, the coefficient for work attitude (WA) is −2.709 in business farms and −1.700 in rice-dominated farms, both of which are significant. Column (1) and (3) show that OLS regressions will greatly underestimate the effect of work attitudes. The coefficient for work attitude is significantly negative, indicating that a serious work attitude will decrease harvest losses (i.e., moral hazards will increase rice harvest losses). Meanwhile, the absolute value of the work attitude coefficient for rice-dominated farms is smaller than that for business farms. Therefore, as the importance of rice production increases, the effects of moral hazards on rice harvest loss rates diminish.

6.4 Robustness Analysis

As in the previous two chapters, the rice advantageous regional dummy variables are also replaced by the rice cropping regional dummy variables (see Chap. 4 for the two specific regional dummy variables).

Table 6.12 gives the estimation results for 651 farms using rice cropping regional controls. The p-value for the endogeneity test is 0.024, which is significant at the $\alpha = 0.05$ level. It indicates the necessity to address the endogeneity of work attitude. The Kleibergen–Paap rk Wald F statistic (Kleibergen and Paap 2006) is 22.541, which is above the 10% critical value (16.38). It means that the instrumental variable is not weak. Therefore, the coefficient of work attitudes estimated by 2SLS regression is more reliable than that estimated by OLS regression, which underestimates the effect, as shown in column (1). The 2SLS regression gives a coefficient of −3.392 for work attitudes (WA), which is significant at the $\alpha = 0.1$ level. This is close to the coefficient of −3.082 obtained using rice advantageous regional controls, which is also significant at the $\alpha = 0.1$ level (see Table 6.6). The coefficients and significance levels of other factors in Table 6.12 are also similar to the results using rice advantageous regional controls. Additionally, the estimations of using rice cropping regional controls for different farming scales, business farms, and rice-dominated farms are also similar to those using rice advantageous regional controls.

Table 6.11 Estimation results of work attitude on HLR (business farms and rice-dominated farms)

Variable	Rice harvest loss rate (%)							
	Business farms				Rice-dominated farms			
	(1) OLS		(2) 2SLS		(3) OLS		(4) 2SLS	
Core independent variable								
WA	−0.403	(0.37)	−2.709*	(1.15)	−0.427	(0.41)	−1.700*	(0.71)
Production and harvesting variable								
Com	−0.956	(0.58)	−0.636	(0.61)	−1.424*	(0.69)	−1.279	(0.67)
Win	0.928	(0.52)	0.802	(0.50)	0.766	(0.51)	0.694	(0.49)
Tra	−0.551	(0.53)	−0.401	(0.51)	0.411	(0.58)	0.505	(0.54)
Wea	1.476*	(0.67)	1.755*	(0.73)	0.410	(0.73)	0.582	(0.76)
Pest = 2	0.234	(0.41)	−0.193	(0.47)	0.039	(0.37)	−0.184	(0.40)
Pest = 3	1.109*	(0.55)	0.620	(0.59)	1.368*	(0.60)	1.058	(0.59)
Area	−0.568	(0.40)	−0.226	(0.45)	−0.477	(0.38)	−0.292	(0.38)
Yield	−0.007	(0.01)	−0.006	(0.01)	−0.009	(0.01)	−0.008	(0.01)
Flat	0.422	(0.62)	0.430	(0.60)	0.492	(0.62)	0.499	(0.60)
Dis	−0.115	(0.29)	−0.281	(0.31)	0.271	(0.25)	0.175	(0.25)
Labor	0.030	(0.45)	−0.054	(0.47)	−0.084	(0.43)	−0.180	(0.43)
Sav	0.336	(0.43)	0.500	(0.43)	0.103	(0.51)	0.216	(0.48)
Mat	−0.065	(0.66)	0.321	(0.68)	0.098	(0.74)	0.291	(0.73)
Price	−4.307**	(1.05)	−4.112**	(0.99)	−2.745**	(1.00)	−2.734**	(1.01)
Household and individual variable								
Gen	0.579	(0.52)	0.406	(0.55)	0.008	(0.49)	−0.103	(0.45)
Age	0.009	(0.02)	0.011	(0.02)	−0.017	(0.02)	−0.018	(0.02)

(continued)

Table 6.11 (continued)

Variable	Rice harvest loss rate (%)							
	Business farms				Rice-dominated farms			
	(1) OLS		(2) 2SLS		(3) OLS		(4) 2SLS	
Edu	−0.161	(0.09)	-0.185^{*}	(0.09)	−0.120	(0.08)	−0.122	(0.08)
Train	0.847	(0.79)	0.955	(0.85)	0.056	(0.82)	0.147	(0.86)
Tinc	0.058	(0.03)	0.047	(0.03)	0.033	(0.04)	0.024	(0.04)
Rincs	-0.017^{*}	(0.01)	-0.019^{*}	(0.01)	-0.017^{*}	(0.01)	-0.018^{**}	(0.01)
Cons	19.241^{**}	(3.76)	18.456^{**}	(3.65)	15.584^{**}	(4.02)	15.466^{**}	(4.00)
Advantageous region	Yes		Yes		Yes		Yes	
R^2	0.306				0.291			
F statistic			27.314				32.071	
Endogeneity test			$p = 0.043$				$p = 0.065$	
N	275		275		215		215	

Notes (1) ***, **, and * indicate 1, 5, and 10% significance levels, respectively. (2) Robust standard errors statistics are in parentheses. (3) Stock–Yogo critical values for weak identification tests (used for Kleibergen–Paap rk Wald F statistics) are 16.38 for 10% maximal IV bias; 8.96 for 15% maximal IV bias; 6.66 for 20% maximal IV bias; and 5.53 for 25% maximal IV bias (Stock and Yogo 2005). (4) The first stage results can be found in columns (2) and (3) of Table A6 in Appendices

Data source Author's calculation based on the survey

Table 6.12 Estimation results of work attitude on HLR (rice cropping regional control)

	Rice harvest loss rate (%)			
	(1) OLS		(2) 2SLS	
Core independent variable				
WA	−0.712**	(0.33)	−3.392***	(1.20)
Production and harvesting variable				
Com	0.307	(0.33)	0.626*	(0.37)
Win	1.155***	(0.26)	1.224***	(0.27)
Tra	−0.561*	(0.30)	−0.443	(0.30)
Wea	0.988*	(0.51)	1.315**	(0.55)
Pest = 2	0.671**	(0.28)	0.556*	(0.30)
Pest = 3	1.217***	(0.36)	0.847**	(0.39)
Area	−0.925***	(0.35)	−0.468	(0.41)
Yield	−0.013**	(0.01)	−0.013**	(0.01)
Flat	0.847**	(0.36)	0.870**	(0.37)
Dis	−0.284	(0.18)	−0.227	(0.18)
Labor	0.765***	(0.29)	0.471	(0.32)
Sav	1.211***	(0.33)	1.327***	(0.33)
Mat	0.026	(0.46)	0.180	(0.49)
Price	−1.543***	(0.58)	−1.871***	(0.60)
Household and individual variable				
Gen	0.701**	(0.31)	0.602*	(0.34)
Age	0.014	(0.01)	0.012	(0.01)
Edu	−0.074	(0.06)	−0.083	(0.06)
Train	1.382***	(0.49)	1.419***	(0.52)
Tinc	0.028	(0.03)	0.023	(0.03)
Rincs	−0.022***	(0.01)	−0.024***	(0.01)
Cons	5.338***	(1.94)	6.319***	(2.05)
Rice cropping region	Yes		Yes	
R^2	0.25			
F statistic			22.541	
Endogeneity test			$p = 0.024$	
N	651		651	

Notes (1) ***, **, and * indicate 1, 5, and 10% significance levels, respectively. (2) Robust standard errors statistics are in parentheses
Data source Author's calculation based on the survey

6.5 Summary

The flourishing harvest outsourcing services has drawn extensive attention from researchers. However, only a few studies have theoretically explored the moral hazard in principal–agent relationship between farmers and market (service providers) and its negative effect on agricultural production. This chapter empirically investigated the effect of moral hazards on rice harvest loss rates. The work attitudes of service providers (whether they treat work seriously) are used to measure moral hazards. Using the data from 651 Chinese households that hired outsourcing services for rice harvesting, the effects of service providers' moral hazards in harvest outsourcing services on rice harvest loss rates are examined from the perspective of farming scale and part-time farming.

Typically, farmers have a more serious work attitude compared to service providers. However, as the farm scale expands, fewer farmers exhibit a diligent work attitude while an increasing number of service providers demonstrate a serious work attitude, leading to a gradual reduction in the gap between them. After dealing with the endogeneity of work attitudes, 2SLS regression results show that moral hazards increase rice harvest loss rates. When the farm scale is small, moral hazard is a crucial factor in increasing rice harvest loss rates. As the farm scale expands, the effect of moral hazards gradually diminishes. The effect of moral hazards on small-scale farms is larger than on large-scale farms. In the largest decile of farms, service providers' work attitudes show no effect on rice harvest losses.

The positive effect of moral hazards on rice harvest loss rates is also observed in business farms. The estimation results given by 2SLS regression show that moral hazards increase rice harvest loss rates. As the importance of rice production increases, the effect of moral hazards on rice harvest loss rates decreases. Rice-dominated farmers, whose biggest source of farming income is rice, are more concerned about rice harvesting than business farmers. The study finds that the effect of moral hazards on rice harvest loss rates is smaller for rice-dominated farms than for business farms.

References

Cai J, Liu WY (2019) Agricultural social service and opportunistic behavior: take agricultural machinery operation services as example. Reform 18–29

Cai LM, Wang LP (2021) Analysis on outsourcing service behavior of rice pest and disease control based on Heckman selection model—a case study of ten counties in Fujian Province. PLoS ONE 16:1–17. https://doi.org/10.1371/journal.pone.0254819

Deng X, De XD, Zeng M, Bin QY (2020) Does outsourcing affect agricultural productivity of farmer households? Evidence from China. China Agric Econ Rev 12:673–688. https://doi.org/10.1108/CAER-12-2018-0236

Dercon S, Gilligan DO, Hoddinott J, Woldehanna T (2009) The impact of roads and agricultural extension on consumption growth and poverty in fifteen Ethiopian villages. Am J Agric Econ 91:1007–1021. https://doi.org/10.1111/j.1467-8276.2009.01325.x

Fan S, Hazell P, Thorat S (2000) Government spending, growth and poverty in rural India. Am J Agric Econ 82:1038–1051. https://doi.org/10.1111/0002-9092.00101

Greeley M (1986) Food, technology and employment: the farm-level post-harvest system in developing countries. J Agric Econ 37:333–347. https://doi.org/10.1111/j.1477-9552.1986.tb01602.x

Huan ML, Hou YX (2020) Quality control contract model of service in agricultural production outsourcing. J Agro-Forestry Econ Manag 19:288–296. https://doi.org/10.16195/j.cnki.cn36-1328/f.2020.03.31

Igata M, Hendriksen A, Heijman W (2008) Agricultural outsourcing: a comparison between the Netherlands and Japan. Appl Stud Agribus Commer 2:29–33. https://doi.org/10.19041/apstract/2008/1-2/4

Ji C, Guo HD, Jin SQ, Yang J (2017) Outsourcing agricultural production: evidence from rice farmers in Zhejiang province. PLoS ONE 12:1–16. https://doi.org/10.1371/journal.pone.0170861

Kaufmann D, Mehrez G, Gurgur T (2019) Voice or public sector management? An empirical investigation of determinants of public sector performance based on a survey of public officials. J Appl Econ 22:321–348. https://doi.org/10.1080/15140326.2019.1627718

Kleibergen F, Paap R (2006) Generalized reduced rank tests using the singular value decomposition. J Econom 133:97–126. https://doi.org/10.1016/J.JECONOM.2005.02.011

Lu QA, Du XD (2020) The outsourcing choice of agricultural production tasks: implications for food security—a multiple-task based approach. In: The 2020 agricultural and applied economics association annual meeting. Kansas City, Missouri

Mi Q, Li XD, Gao JZ (2020) How to improve the welfare of smallholders through agricultural production outsourcing: evidence from cotton farmers in Xinjiang, Northwest China. J Clean Prod 256:120636. https://doi.org/10.1016/j.jclepro.2020.120636

Picazo-Tadeo AJ, Reig-Martínez E (2006) Outsourcing and efficiency: the case of Spanish citrus farming. Agric Econ 35:213–222. https://doi.org/10.1111/j.1574-0862.2006.00154.x

Stock JH, Yogo M (2005) Testing for weak instruments in linear IV regression. In: Identification and inference for econometric models: essays in honor of Thomas Rothenberg. Cambridge University Press, Cambridge

Wuepper D, Yesigat Ayenew H, Sauer J (2018) Social capital, income diversification and climate change adaptation: panel data evidence from rural Ethiopia. J Agric Econ 69:458–475. https://doi.org/10.1111/1477-9552.12237

Yang J, Huang ZH, Zhang XB, Reardon T (2013) The rapid rise of cross-regional agricultural mechanization services in China. Am J Agric Econ 95:1245–1251. https://doi.org/10.1093/ajae/aat027

Yi Q (2018) Adoption of agricultural mechanization services among maize farmers in China: impacts of population aging and off-farm employment. In: 30th international conference of agricultural economists. Vancouver

Zhang XB, Yang J, Thomas R (2017) Mechanization outsourcing clusters and division of labor in Chinese agriculture. China Econ Rev 43:184–195. https://doi.org/10.1016/j.chieco.2017.01.012

Chapter 7
Conclusion and Policy Implications

In the recent years, there is diminishing room to increase food supply by increasing food production due to population growth, scarcity of natural resources, and climate change (Shafiee-Jood and Cai 2016). An alternative approach to increasing food supply—reducing food loss and waste—has gained significant attention as it can increase food availability without investing additional resource inputs, such as land and water.

Food loss and waste happen throughout the entire process from field to fork, involving many participants along the food supply chain, such as farmers, intermediaries, and retailers. In less developed countries and regions, more food is lost at the front end of the food supply chain than that wasted at the end of the food supply chain due to inadequate harvesting techniques, poor infrastructure, lack of storage facilities, and inadequate skills and management (FAO 2013; Lipinski et al. 2013; Neff et al. 2015). Food loss reduces farmers' income and threatens their food security, especially for farmers in less developed countries and regions (Coker and Ninalowo 2016; Danbaba et al. 2019). Furthermore, harvest losses account for the highest proportion of food loss suffered by farmers (Basappa et al. 2007) and most farmers suffer losses during harvesting (Ibrahim et al. 2018).

Among the various factors influencing harvest losses, machinery is a crucial factor. Studies have examined the role of machinery in harvest losses (Benaseer et al. 2018; Huang et al. 2018; Kantor et al. 1997; Parfitt et al. 2010), but ignored the outsourcing services involved, which have significantly contributed to the development of mechanical harvesting in China (Sheng et al. 2017). Some studies have pointed out due to asymmetric information and inconsistent goals, service providers of harvest outsourcing services exhibit moral hazard behaviors, such as increasing the forward speed of machines and reducing efforts, which may negatively affect agriculture production (Cai and Liu 2019; Chegere 2017; Coelli and Battese 1996; Huan and Hou 2020; Sun 2013). However, there is a lack of empirical studies on the moral hazard in harvest outsourcing services and its effect on agriculture.

X. Qu et al., *Harvest Loss in China*, The University of Tokyo Studies on Asia,
https://doi.org/10.1007/978-981-97-9156-9_7

Therefore, the objectives of this study are to quantify rice harvest losses in China and to explore the effects of harvest outsourcing services on rice harvest losses through moral hazards. Specifically, the following research objectives are included: (1) to estimate rice harvest losses in China; (2) to investigate moral hazards in harvest outsourcing services; (3) to explore the effect of harvest outsourcing services on rice harvest losses and to determine whether this effect mediated by achieved through moral hazards. (4) To analyze the effect of moral hazards in harvest outsourcing services on rice harvest losses.

7.1 Conclusion

First, the rice harvest loss rate was 3.65% in China, equivalent to nearly 8 million tons of rice and more than 1 million hectares of farmland. The harvest loss rates varied by farming scale, region, harvesting method, and operator. The average rice harvest loss rates decreased with the expansion of the farming scale. The highest rice harvest loss rate was found in the Southeast Coast (4.49%), followed by the Yangtze River Basin (3.52%) and the Northeast Plains (2.87%). The harvest loss rate for segmented harvesting (3.88%) was higher than that for combine harvesting (3.39%). More than half of the farmers purchased harvest outsourcing services (58.86%) and their average rice harvest loss rate was 3.50%, slightly lower than that of farmers who did not purchase outsourcing services.

Second, service providers have moral hazards. The work attitude during harvesting is used to capture the effort level of operators. Moral hazard is considered to exist if the service provider is less serious than the farmers. Both descriptive analysis and Logit regressions show that service providers' work attitude is less serious than that of the farmers. Since the service fee for combine harvesting is higher than that for segmented harvesting, combine harvesting can alleviate service providers' moral hazard through income incentives. Therefore, in small-scale farms, large-scale farms, and business farms, the less serious work attitude of service providers is only found in segmented harvesting. Robustness analyses are conducted by PSM and by replacing rice advantageous regions with rice cropping regions. They all indicate that service providers have less serious work attitudes than farmers, thus confirming the existence of moral hazards.

Third, purchasing harvest outsourcing services will increase rice harvest loss rates, in which moral hazard mediates this positive effect. Mediation analysis models are applied to test the hypothesis that moral hazards—measured by whether the work attitudes of harvest operators are serious—mediate the effect of harvest outsourcing services on rice harvest loss rates. The marginal effect of harvest outsourcing services on rice harvest loss rates given by Tobit regression is 0.393, which is significant at the $\alpha = 0.1$ level, and it decreases to 0.132 and becomes insignificant in IV-Tobit regression after the mediator is included as an independent variable. Moreover, the marginal effect of outsourcing services on work attitudes is significant. Therefore, the three-step method indicates the existence of mediation effect. Meanwhile, the adjusted product of coefficients method with the Sobel test, the distribution of the product test, and the MCMC test further indicate the significance of mediation effect. This positive effect of outsourcing services on harvest loss rates and the mediation effect of moral hazards is also found in large-scale farms. However, when applying combine harvesting, outsourcing services have no effect on harvest loss rates and the mediation effect of moral hazard does not exist. Similar results are obtained by changing the rice regional controls.

Fourth, service providers' moral hazards will increase the rice harvest loss rates and this effect decreases as farm scale expands and the importance of rice production increases. We directly analyze the effect of service providers' moral hazards on rice harvest loss rates of farms using harvest outsourcing services. After addressing the endogeneity of moral hazards, 2SLS regression results show that moral hazards increase rice harvest loss rates. As farm size expands, the effect of service providers' moral hazards on rice harvest loss rates decreases. This positive effect of moral hazards is also observed among business farms and rice-dominated farms. It becomes smaller as the importance of rice production increases.

7.2 Policy Implication

Although this study examines the moral hazard problem in harvest outsourcing services and its impact on rice harvest losses, it is not intended to oppose outsourcing services. Harvest outsourcing services enable farmers who cannot afford machines to harvest rice in time. Thus, from this perspective, outsourcing services can reduce harvest losses. However, this is different from the perspective of moral hazard and goes beyond the scope of this study. While a harvest loss rate of 3.65% may seem small, the total absolute amount is staggering in China. As a country with a large population and a large food importer, the importance of China's food supply cannot be overstated. Moreover, food loss is not only an economic loss, but also an environmental and resource loss. Therefore, although harvest loss occurs on the farm, loss reduction is more of a social issue for which society and governments should be held accountable. Based on the findings in this study, several policy recommendations are proposed mainly from the perspective of reducing moral hazard.

First, promoting combine harvest outsourcing services. The research results suggest that combine harvest outsourcing services can reduce rice harvest losses, especially for farmers with relatively large planting areas. Therefore, the government should continue to implement subsidies for combine harvesters and expand harvest outsourcing service team. Moreover, the government should strengthen the organization and dispatch of harvest outsourcing services. Guiding agricultural machinery specialized service households and organizations to provide harvesting service in an orderly manner from south to north, so that more farmers can get access to outsourcing services for combine harvesting in time during the busy season.

Second, promoting the transformation and upgrading of agricultural equipment. The results show that existing combine harvesters are not suitable for harvesting on small farmland. The government should strengthen agricultural machinery research and development and its promotion. Research institutions and manufacturing enterprises should be encouraged and supported to develop high-efficiency, low-loss harvesting machines, and accelerate the promotion and application of new technologies and equipment.

Third, improving the technical management level of agricultural machinery operators. The results suggest that it is essential to address the moral hazards of service providers to reduce harvest losses. Strengthening mechanical operation management and training to promote standardized harvesting operations. Increasing awareness among both operators and farmers about grain conservation and loss reduction. The government should strengthen technical guidance for farmers and operators, guiding them to choose the suitable equipment and optimal harvesting periods to increase service quality and avoid increased losses.

Fourth, regulating the outsourcing service market for agricultural machinery. Encouraging farmers and service providers to sign standardized written service contracts. The written contract should provide detailed provisions on various aspects, such as operation content, operation standard and settlement, rights and obligations of both parties, liability for breach of contract, and mediation in case of disputes in harvest outsourcing services. In addition, it is necessary to clarify the service fees for harvest outsourcing services and how they are calculated. Currently, service fee is proportional to the serviced area. The irresponsibility of service providers for the quality of harvesting is an important cause of moral hazards. A reasonable service fee calculation method that considers the responsibilities of the service provider should be introduced, such as basing fees on the harvested yield.

Fifth, guiding the orderly transfer of rural land management rights for the development of agricultural operations at an appropriate scale. The results indicate that an appropriate scale is crucial for mitigating the impact of moral hazards on rice harvest losses. Although there is an ongoing promotion of outsourcing services in China to help small farms get access to agricultural machinery, it is undeniable that machinery, especially the combine harvester, is more suitable for operating on large farmland. The government should encourage local communities to establish land management rights transfer markets to standardize services related to policy consultation and information dissemination for land management rights transfer.

Sixth, strengthening agricultural meteorological services and pest and disease control. The government should promptly issue severe weather warnings to remind farmers to harvest rice in advance of impending severe weather, particularly typhoon and storm warnings in the Southeast coast. Conducting artificial weather modification operations to minimize harvest losses due to meteorological disasters. Strengthening the construction and management of crop pest and disease monitoring network to standardize the monitoring content and information reporting. Encouraging and supporting specialized pest control service organizations by providing them with technical training, guidance, and services.

7.3 Contribution and Future Study

This study makes three contributions to the literature. First, it provides the first nationwide household survey data on rice harvest losses since 1995, offering the most recent nationwide data available. Second, to our knowledge, this is the first study in China to examine the effect of harvest outsourcing services on rice harvest losses, providing a new perspective for evaluating the effect of outsourcing services on agricultural production. Third, using mediation analysis models, this study empirically verifies for the first time that moral hazards mediate the effect of harvest outsourcing services on rice harvest losses, thus providing empirical evidence for the existence of moral hazards in harvest outsourcing services.

Nonetheless, the study has several limitations, and future research could address them in the following ways. First, studying the potential for reducing harvest losses, intervention measures, and costs and benefits. Calculating the loss reduction potential for different intervention measures and analyzing their costs and benefits, including environmental and resource benefits, to determine the optimal harvest loss rate. This is essential for promoting the implementation of interventions and clarifying the responsible parties for their execution. Second, measuring quality losses during harvesting. The measurement of grain quality losses is more complex than the quantitative losses. Before rice reaches formal market, farmers lack sufficient quality awareness to distinguish between different grades of rice (Hodges et al. 2014). Studying grain quality losses is challenging, but meaningful. Third, studying harvest losses in different regions. The characteristics of rice production vary by region, leading to different key factors affecting harvest losses. For example, weather significantly affects harvest losses in the Southeast Coast. Future in-depth studies could be conducted in regions with similar characteristics to develop more targeted interventions. Fourth, studying the moral hazard issue in outsourcing services in other agricultural production processes. Due to the special attributes of agricultural production, coupled with the expanding scope of agricultural outsourcing services and the deepening marketization of agricultural production in the future, moral hazard problem in agricultural outsourcing services will be more prominent and complex than that in the secondary and tertiary industries, especially in technology-intensive stages of production, and therefore more research is required.

References

Basappa G, Deshmanya JB, Patil BL (2007) Post-harvest losses of maize crop in Karnataka—an economic analysis. Karnataka J Agric Sci 20:69–71

Benaseer S, Masilamani P, Albert VA et al (2018) Harvesting and threshing methods their impact on seed quality: a review. Agric Rev 39:183–192. https://doi.org/10.18805/ag.R-1803

Cai J, Liu WY (2019) Agricultural social service and opportunistic behavior: take agricultural machinery operation services as example. Reform 18–29

Chegere MJ (2017) Post-harvest losses, intimate partner violence and food security in Tanzania. University of Gothenburg, Gothenburg, Sweden

Coelli T, Battese G (1996) Identification of factors which influence the technical inefficiency of Indian farmers. Aust J Agric Econ 40:103–128. https://doi.org/10.1111/j.1467-8489.1996.tb00558.x

Coker AA, Ninalowo SO (2016) Effect of post-harvest losses on rice farmers' income in Sub-saharan Africa: a case of Niger state, Nigeria. J Agric Sci Food Technol 2:27–34

Danbaba N, Idakwo PY, Kassum AL et al (2019) Rice postharvest technology in Nigeria: an overview of vurrent status, constraints and potentials for sustainable development. Open Access Libr J 6:1–23. https://doi.org/10.4236/oalib.1105509

FAO (2013) Food wastage footprint: Impacts on natural resources. Food and Agriculture Organization of the United Nations, Rome, Italy

Hodges RJ, Bernard M, Rembold F (2014) APHLIS—postharvest cereal losses in Sub-Saharan Africa, their estimation, assessment and reduction. European Union, Luxembourg

Huan ML, Hou YX (2020) Quality control contract model of service in agricultural production outsourcing. J Agro-Forestry Econ Manag 19:288–296. https://doi.org/10.16195/j.cnki.cn36-1328/f.2020.03.31

Huang D, Yao L, Wu LP, Zhu X Di (2018) Measuring rice loss during harvest in China: based on experiment and survey in five provinces. J Nat Resour 33:1427–1438. https://doi.org/10.31497/zrzyxb.20170810

Ibrahim H, Saba S, Ojoko EA (2018) Post-harvest loss in rice production: evidence from a rural community in Northern Nigeria. FUDMA J Sci 2:17–22

Kantor LS, Lipton K, Manchester A, Oliveira V (1997) Estimating and addressing America's food losses. Food Rev Food Rev 20:2–12. https://doi.org/10.22004/ag.econ.234453

Lipinski B, Hanson C, Lomax J et al (2013) Reducing food loss and waste. World Resources Institute, Washington, D.C.

Neff RA, Kanter R, Vandevijvere S (2015) Reducing food loss and waste while improving the public's health. Health Aff 34:1821–1829. https://doi.org/10.1377/hlthaff.2015.0647

Parfitt J, Barthel M, MacNaughton S (2010) Food waste within food supply chains: quantification and potential for change to 2050. Philos Trans R Soc B Biol Sci 365:3065–3081. https://doi.org/10.1098/rstb.2010.0126

Shafiee-Jood M, Cai X (2016) Reducing food loss and waste to enhance food security and environmental sustainability. Environ Sci Technol 50:8432–8443. https://doi.org/10.1021/acs.est.6b01993

Sheng Y, Song LG, Yi Q (2017) Mechanisation outsourcing and agricultural productivity for small farms: implications for rural land reform in China. In: Song L, Garnaut R, Cai F, Johnston L (eds) China's new sources of economic growth: human capital, innovation and technological change, vol. 2. Australian National University, Acton, Australia, pp 289–313

Sun XH (2013) Agricultural production operator: comparison of types and path choice—from the ferspective of the overall production efficiency. Res Econ Manag 59–66. https://doi.org/10.13502/j.cnki.issn1000-7636.2013.12.007

Appendices

See Tables A.1, A.2, A.3, A.4, A.5, A.6 and A.7.

Table A.1 Mediation effect test: the adjusted product of coefficients method (1106 farms)

Variable	HLR				WA	
	(1) OLS		(2) 2SLS		(3) Logit	
Core independent variable						
OS	0.364	(0.32)	-1.309^{*}	(0.72)	-2.295^{***}	(0.31)
Mediator						
WA			-5.540^{***}	(2.04)		
Production and harvesting variable						
Com	0.149	(0.29)	0.721^{*}	(0.39)	0.969^{***}	(0.28)
Win	1.016^{***}	(0.21)	1.227^{***}	(0.25)		
Tra	-0.984^{***}	(0.24)	-1.292^{***}	(0.29)		
Wea	1.242^{***}	(0.38)	0.742^{*}	(0.44)	-0.630^{**}	(0.28)
Pest = 2	0.823^{***}	(0.22)	0.117	(0.37)	-0.724^{***}	(0.19)
Pest = 3	2.277^{***}	(0.31)	1.614^{***}	(0.42)	-0.774^{***}	(0.23)
Area	-1.685^{***}	(0.40)	−0.348	(0.63)	1.486^{***}	(0.33)
Yield	-0.010^{**}	(0.00)	-0.012^{**}	(0.00)		
Flat	0.016	(0.27)	0.589^{*}	(0.36)	0.702^{***}	(0.23)
Dis	−0.148	(0.12)	−0.189	(0.13)		
Labor	0.597^{**}	(0.23)	0.411	(0.27)	−0.152	(0.17)
Sav	0.753^{***}	(0.26)	1.123^{***}	(0.32)	0.383^{*}	(0.22)
Mat	0.253	(0.44)	0.503	(0.53)		

(continued)

X. Qu et al., *Harvest Loss in China*, The University of Tokyo Studies on Asia,
https://doi.org/10.1007/978-981-97-9156-9

Table A.1 (continued)

Variable	HLR				WA	
	(1) OLS		(2) 2SLS		(3) Logit	
Price	−0.411	(0.35)	0.005	(0.43)	0.200	(0.24)
Household and individual variable						
Gen	−0.102	(0.26)	−0.060	(0.31)	0.091	(0.23)
Age	0.024**	(0.01)	0.035***	(0.01)	0.015*	(0.01)
Edu	−0.006	(0.05)	0.038	(0.05)	0.042	(0.03)
Train	0.642*	(0.37)	0.488	(0.40)	−0.166	(0.31)
Tinc	0.013	(0.02)	−0.015	(0.02)	−0.036*	(0.02)
Rincs	−0.004	(0.01)	−0.011	(0.01)	−0.014**	(0.01)
Htor					0.320***	(0.10)
Cons	3.809***	(1.47)	2.765	(1.77)	−3.315***	(1.02)
Advantageous region	Yes		Yes		Yes	
N	1106		1106		1106	
Pseudo R^2/R^2	0.188				0.147	

Notes (1) Robust standard errors are reported in parentheses. (2) Significance levels are as follows: *** = 1%, ** = 5%, and * = 10%. (3) All regressions are based on the data after winsorizing
Data source Author's calculation based on the survey

Table A.2 Mediation effect test: the adjusted product of coefficients method (Large-scale farms)

Variable	HLR				WA	
	(1) OLS		(2) 2SLS		(3) Logit	
Core independent variable						
OS	1.126***	(0.42)	0.213	(0.60)	−2.371***	(0.45)
Mediator						
WA			−2.789**	(1.20)		
Production and harvesting variable						
Com	−0.229	(0.34)	0.299	(0.45)	1.260***	(0.34)
Win	0.754***	(0.29)	0.916***	(0.30)		
Tra	−0.453	(0.33)	−0.516	(0.35)		
Wea	0.984*	(0.54)	1.133**	(0.56)	0.512	(0.41)
Pest = 2	0.694***	(0.27)	0.329	(0.33)	−0.667**	(0.29)
Pest = 3	1.522***	(0.37)	1.027**	(0.45)	−1.056***	(0.32)
Area	−0.590*	(0.33)	−0.128	(0.39)	0.876**	(0.34)
Yield	−0.010*	(0.01)	−0.010*	(0.01)		

(continued)

Table A.2 (continued)

Variable	HLR				WA	
	(1) OLS		(2) 2SLS		(3) Logit	
Flat	0.471	(0.34)	0.404	(0.35)	−0.194	(0.31)
Dis	−0.086	(0.13)	−0.120	(0.12)		
Labor	0.548*	(0.28)	0.363	(0.31)	−0.463	(0.32)
Sav	0.266	(0.30)	0.420	(0.32)	0.194	(0.34)
Mat	0.060	(0.49)	0.396	(0.50)		
Price	−1.113	(0.77)	−1.791**	(0.85)	−2.034***	(0.77)
Household and individual variable						
Gen	−0.157	(0.32)	−0.123	(0.33)	0.377	(0.35)
Age	0.017	(0.01)	0.024	(0.01)	0.014	(0.01)
Edu	−0.058	(0.06)	−0.052	(0.06)	0.007	(0.05)
Train	0.172	(0.47)	0.255	(0.52)	0.079	(0.48)
Tinc	0.055**	(0.03)	0.047*	(0.03)	−0.024	(0.02)
Rincs	−0.013*	(0.01)	−0.015*	(0.01)	−0.009	(0.01)
Htor					0.728***	(0.19)
Cons	6.039**	(2.56)	8.117***	(2.84)	4.275*	(2.49)
Advantageous region	Yes		Yes		Yes	
N	532		532		532	
Pseudo R^2/R^2	0.152				0.157	

Notes (1) Robust standard errors are reported in parentheses. (2) Significance levels are as follows: *** = 1%, ** = 5%, and * = 10%. (3) All regressions are based on the large-scale sample after winsorizing

Data source Author's calculation based on the survey

Table A.3 Mediation effect test: the adjusted product of coefficients method (Large-scale farms using combines)

Variable	HLR				WA	
	(1) OLS		(2) 2SLS		(3) Logit	
Core independent variable						
OS	0.450	(0.49)	0.000	(0.59)	−0.984	(0.79)
Mediator						
WA			−2.005**	(0.91)		
Production and harvesting variable						
Win	1.311***	(0.40)	1.622***	(0.37)		
Tra	−0.541	(0.45)	−0.408	(0.45)		
Wea	−0.052	(0.59)	0.145	(0.59)	0.796*	(0.48)

(continued)

Table A.3 (continued)

Variable	HLR				WA	
	(1) OLS		(2) 2SLS		(3) Logit	
Pest = 2	0.574	(0.35)	0.498	(0.35)	0.183	(0.44)
Pest = 3	0.675	(0.49)	0.195	(0.49)	−1.255***	(0.47)
Area	−0.363	(0.35)	−0.017	(0.34)	1.056**	(0.49)
Yield	−0.007	(0.01)	−0.011	(0.01)		
Flat	0.890*	(0.49)	1.005**	(0.50)	−0.265	(0.47)
Dis	−0.012	(0.22)	−0.024	(0.21)		
Labor	−0.197	(0.32)	−0.539	(0.36)	−1.317**	(0.56)
Sav	−0.204	(0.41)	−0.266	(0.40)	−0.395	(0.57)
Mat	−0.832	(0.52)	−0.637	(0.52)		
Price	−1.164	(0.94)	−1.297	(0.92)	−1.148	(1.13)
Household and individual variable						
Gen	0.368	(0.34)	0.282	(0.35)	0.062	(0.43)
Age	0.017	(0.02)	0.013	(0.02)	−0.019	(0.02)
Edu	−0.058	(0.07)	−0.068	(0.07)	−0.015	(0.07)
Train	0.283	(0.59)	0.003	(0.60)	−1.831	(1.22)
Tinc	0.032	(0.03)	0.030	(0.03)	−0.016	(0.02)
Rincs	−0.023***	(0.01)	−0.024***	(0.01)	−0.013	(0.01)
Htor					0.978***	(0.24)
Cons	7.192**	(2.78)	8.196***	(2.77)	2.194	(3.62)
Advantageous region	Yes		Yes		Yes	
N	302		302		302	
Pseudo R^2/R^2	0.218				0.220	

Notes (1) Robust standard errors are reported in parentheses. (2) Significance levels are as follows: *** = 1%, ** = 5%, and * = 10%. (3) All regressions are based on the large-scale sample after winsorizing

Data source Author's calculation based on the survey

Table A.4 Estimation results of first stage

Variable	Work attitude							
	(1) 1106 farms		(2) 1106 farms		(3) Large-scale farms		(4) Large-scale farms using combines	
Htor	0.065***	(0.02)	0.077***	(0.02)	0.157***	(0.04)	0.194***	(0.05)
OS	−0.302***	(0.04)	−0.174***	(0.03)	−0.338***	(0.06)	−0.217*	(0.13)
Com	0.101***	(0.03)			0.181***	(0.04)		
Win	0.054**	(0.03)			0.099***	(0.04)	0.212***	(0.06)
Tra	−0.052*	(0.03)			−0.010	(0.04)	0.077	(0.06)
Wea	−0.075**	(0.04)			0.095*	(0.06)	0.174**	(0.08)
Pest = 2	−0.116***	(0.03)			−0.090**	(0.04)	0.021	(0.06)
Pest = 3	−0.122***	(0.03)			−0.156***	(0.05)	−0.202***	(0.06)
Area	0.250***	(0.05)			0.166***	(0.06)	0.149**	(0.07)
Yield	−0.000	(0.00)			0.000	(0.00)	−0.001	(0.00)
Flat	0.102***	(0.03)			−0.041	(0.04)	0.011	(0.07)
Dis	−0.022	(0.02)			−0.005	(0.03)	0.013	(0.05)
Labor	−0.029	(0.03)			−0.060	(0.04)	−0.152***	(0.06)
Sav	0.058*	(0.03)			0.036	(0.05)	−0.049	(0.08)
Mat	0.045	(0.05)			0.114*	(0.06)	0.097	(0.08)
Price	0.040	(0.05)			−0.312***	(0.01)	−0.267	(0.19)
Gen	0.003	(0.04)			0.011	(0.05)	−0.044	(0.07)
Age	0.002	(0.00)			0.002	(0.00)	−0.002	(0.00)
Edu	0.008	(0.01)			0.002	(0.01)	−0.006	(0.01)
Train	−0.027	(0.04)			0.040	(0.07)	−0.161*	(0.10)

(continued)

Table A.4 (continued)

Variable	Work attitude							
	(1) 1106 farms		(2) 1106 farms		(3) Large-scale farms		(4) Large-scale farms using combines	
Tinc	−0.005**	(0.00)			−0.005*	(0.00)	−0.004	(0.00)
Rincs	−0.002	(0.00)			−0.001	(0.00)	−0.001	(0.00)
Cons	−0.100	(0.19)	0.251**	(0.04)	0.880***	(0.33)	1.094*	(0.61)
Advantageous region	Yes		Yes		Yes		Yes	
N	1106		1106		532		302	
Kleibergen–Paap *F* statistic	11.506		15.920		14.535		17.108	

Notes (1) Robust standard errors are reported in parentheses. (2) Significance levels are as follows: *** = 1%, ** = 5%, and * = 10%. (3) The coefficients and Kleibergen–Paap *F* statistics are the same as those in 2SLS

Data source Author's calculation based on the survey

Table A.5 Estimation results of first stage (farming scale)

Variable	Work attitude					
	(1) Small-scale farms		(2) Large-scale farms		(3) Largest decile of farms	
Htor	0.117***	(0.03)	0.239***	(0.05)	0.565***	(0.10)
Com	0.052	(0.04)	0.187***	(0.04)	0.109	(0.18)
Win	−0.005	(0.04)	0.159***	(0.05)	0.037	(0.29)
Tra	0.056	(0.04)	0.032	(0.06)	0.221	(0.43)
Wea	0.020	(0.06)	0.123*	(0.07)	−0.137	(0.27)
Pest = 2	−0.059	(0.04)	−0.005	(0.06)	−0.150	(0.14)
Pest = 3	−0.078	(0.05)	−0.193***	(0.06)	−0.078	(0.18)
Area	0.139	(0.26)	0.128*	(0.07)	−0.254**	(0.13)
Yield	0.001	(0.00)	0.000	(0.00)	0.004	(0.00)
Flat	0.075	(0.05)	−0.027	(0.06)	0.069	(0.31)
Dis	0.030	(0.05)	−0.006	(0.04)	0.082	(0.13)
Labor	−0.107***	(0.03)	−0.057	(0.06)	0.078	(0.14)
Sav	0.063	(0.06)	−0.020	(0.07)	0.421**	(0.16)
Mat	0.057	(0.09)	0.137*	(0.08)	0.255	(0.22)
Price	−0.295***	(0.11)	−0.043	(0.14)	−0.837	(0.56)
Gen	−0.049	(0.06)	−0.003	(0.08)	0.185	(0.13)
Age	0.000	(0.00)	−0.001	(0.00)	−0.004	(0.01)
Edu	−0.006	(0.01)	−0.005	(0.01)	−0.015	(0.02)
Train	0.082	(0.05)	−0.025	(0.10)	0.084	(0.22)
Tinc	−0.006	(0.00)	−0.010***	(0.00)	−0.027*	(0.02)
Rincs	−0.003*	(0.00)	−0.002	(0.00)	0.003	(0.00)
Cons	0.687*	(0.36)	0.018	(0.44)	2.019	(2.09)
Advantageous region	Yes		Yes		Yes	
N	329		322		65	
Kleibergen–Paap *F* statistic	11.745		23.467		30.452	

Notes (1) Robust standard errors are reported in parentheses. (2) Significance levels are as follows: *** = 1%, ** = 5%, and * = 10%

Data source Author's calculation based on the survey

Table A.6 Estimation results of first stage (business farms)

Variable	Work attitude					
	(1) 651 farms		(2) Business farms		(3) Rice-dominated farms	
Htor	0.145***	(0.03)	0.184***	(0.04)	0.248***	(0.04)
Com	0.132***	(0.03)	0.106**	(0.05)	0.064	(0.06)
Win	0.072**	(0.03)	−0.016	(0.05)	−0.020	(0.06)
Tra	0.069**	(0.03)	0.109**	(0.05)	0.130**	(0.07)
Wea	0.110**	(0.05)	0.187**	(0.08)	0.180*	(0.10)
Pest = 2	−0.033	(0.04)	−0.108**	(0.05)	−0.096	(0.06)
Pest = 3	−0.131***	(0.04)	−0.170***	(0.06)	−0.205***	(0.06)
Area	0.190***	(0.06)	0.131**	(0.06)	0.108	(0.07)
Yield	0.000	(0.00)	0.001	(0.00)	0.001	(0.00)
Flat	−0.006	(0.04)	−0.045	(0.06)	−0.028	(0.07)
Dis	0.009	(0.03)	−0.082**	(0.04)	−0.103**	(0.05)
Labor	−0.095***	(0.03)	−0.053	(0.05)	−0.056	(0.06)
Sav	0.019	(0.04)	0.098	(0.07)	0.152*	(0.09)
Mat	0.056	(0.06)	0.172**	(0.08)	0.081	(0.11)
Price	−0.243***	(0.08)	−0.215	(0.15)	−0.283*	(0.16)
Gen	−0.039	(0.05)	−0.091	(0.08)	−0.092	(0.10)
Age	−0.001	(0.00)	−0.001	(0.00)	−0.003	(0.00)
Edu	−0.004	(0.01)	−0.013	(0.01)	−0.004	(0.01)
Train	0.012	(0.05)	0.076	(0.10)	0.111	(0.14)
Tinc	−0.006**	(0.00)	−0.008***	(0.00)	−0.009***	(0.00)
Rincs	−0.001	(0.00)	−0.002	(0.00)	−0.002	(0.00)
Cons	0.592**	(0.29)	0.693	(0.51)	1.038	(0.58)
Advantageous region	Yes		Yes		Yes	
N	651		275		215	
Kleibergen–Paap *F* statistic	25.311		27.314		32.071	

Notes (1) Robust standard errors are reported in parentheses. (2) Significance levels are as follows: *** = 1%, ** = 5%, and * = 10%

Data source Author's calculation based on the survey

Table A.7 Estimations of rice harvest losses in literature

Method	Data year	Region(s)	Magnitude	Citation(s)
FLA methodology	–	Democratic Republic of Congo	Harvesting: 12% Shelling/Threshing/Dehulling: 11%	Totobesola et al. (2022)
Field experiments	–	Egypt	Manual reaping and tractor threshing: 2.49% Manual reaping and local thresher: 2.03% Combine harvesting: 1.35%	Badawi (2001)
Questionnaire	–	Ghana	Farmers' perceptions about harvesting loss 53.7% of farmers thought: 0–9% 36.11% of farmers thought: 10–19% 10.19% of farmers thought: 20–29%	Amponsah et al. (2018)

(continued)

Table A.7 (continued)

Method	Data year	Region(s)	Magnitude	Citation(s)
Field experiments, questionnaire		Ghana	Farmers' perceptions about the total post-harvest loss (from harvesting to milling): 35% of farmers thought: over 40% 35% of farmers thought: 30–39% 15% of farmers thought: 20–29% 10% of farmers thought: 10–19% 5% of farmers thought: 0–9% Harvesting loss at 4*5-m area of rice field experiments: Nerica 1 by panicle: 6450 g (1.13%) Nerica 2 by panicle: 6409 g (1.64%) Average loss of Nerica 1 and 2 by panicle: 6430 g (1.38%) Nerica 1 by sickle: 6925 g (3.25%) Nerica 2 by sickle: 7443 g (2.62%) Average loss of Nerica 1 and 2 by sickle: 7184 g (2.93%) Average loss of Nerica 1: 6688 g (2.10%) Average loss of Nerica 2: 6926 g (2.33%) Threshing loss at 4*5-m area of rice field experiments: Nerica 1 by bag-beating (panicle): 3.98% Nerica 2 by bag-beating (panicle): 0.92% Average loss of Nerica 1 and 2 by bag-beating (panicle): 2.45% Nerica 1 by Bambam (sickle): 5.33% Nerica 2 by Bambam (sickle): 6.96% Average loss of Nerica 1 and 2 by Bambam (sickle): 6.14% Average loss of Nerica 1: 4.65% Average loss of Nerica 2: 3.94% Harvesting and threshing loss at farmers' fields: Farmer 1 (Nerica): harvesting loss (Sickle): 382 g (7.91%); threshing loss (Sac beating): 35 g (0.73%); total loss: 8.65% Farmer 2 (Nerica): harvesting loss (Sickle): 135 g (12.05%); threshing loss (Sac beating): 50 g (4.07%); total loss: 16.14% Farmer 3 (Nerica): harvesting loss (Sickle): 198 g (2.60%); threshing loss (Sac beating): 211 g (3.00%); total loss: 5.60% Farmer 4 (Sikamo): harvesting loss (Sickle): 299 g (8.20%); threshing loss (Sac beating): 144 g (3.73%); total loss: 11.93% Farmer 5 (Nerica): harvesting loss (Sickle): 177 g (3.03%); threshing loss (Sac beating): 36 g (0.53%); total loss: 3.57%	Guisse (2010), Appiah et al. (2011)

(continued)

Table A.7 (continued)

Method	Data year	Region(s)	Magnitude	Citation(s)
Field experiments		Nobewam (Ghana)	Qualitative threshing loss: Variety 1: by Bambam (14.3%); by drum (13.8%); by sack (12.8%) Variety 2: by Bambam (12.8%); by drum (15.5%); by sack (10.0%) Average loss of variety 1 and 2: by Bambam (13.49%); by drum (14.65%); by sack (11.37%) Economic threshing loss (GH¢): Variety 1: by Bambam (81.2); by drum (78.7); by sack (72.6) Variety 2: by Bambam (72.6); by drum (88.4); by sack (57.0) Average loss of variety 1 and 2: by Bambam (76.86); by drum (83.51); by sack (64.79)	Sanneh (2015)
Field experiments	2014–2015	Nigeria	Bag-beating: reaping (1.56%), threshing (2.27%), winnowing (1.01%) Bambam: reaping (1.77%), threshing (4.15%), winnowing (1.26%) Machinery: reaping (1.90%), threshing (5.96%), winnowing (1.47%)	Amusat et al. (2016)
Field experiments, questionnaire	–	Nigeria	Reaping: 4.42%, threshing and winnowing: 4.97%, transportation from field to home: 0.34% Danbaba et al. (2019) used the above estimation and the 2016 paddy production in Nigeria to calculate the corresponding rice quantity and quality losses: Reaping: 0.78 million metric tons; 104.66 billion naira Threshing and winnowing: 0.87 million metric tons; 117.42 billion naira Transportation from field to home: 0.60 million metric tons; 8.03 billion naira	Oguntade et al. (2014), Danbaba et al. (2019)
Questionnaire	2014	Nigeria	Reaping: 0.15 kg per farmer, threshing: 0.25 kg per farmer, winnowing: 0.15 kg per farmer	Coker and Ninalowo (2016)
Field experiments	2018	Sub-Sahara Africa	Shattering loss during reaping: 2.8% Stacking loss after reaping and before threshing: 4.2% Manual threshing: unthreshed loss (1.9% ± 1.3%), scattered loss (1.6% ± 1.3%)	Ndindeng et al. (2021)
Field experiments, face to face interview	2010	Uganda	Harvesting loss: 6.77% (285.1 kg/ha) Threshing loss: 4.7% (183.0 kg/ha) Winnowing loss: 1.03% (34.6 kg/ha)–1.63% (53.4 kg/ha)	Candia et al. (2012)
Field experiments	1979–1980	Bangladesh	Threshing loss: By bullock treading: 2.54% By hand beating and bullock treading: short straw (0.60%), long straw (1.45%) By pedal thresher: short straw (1.82%), long straw (3.49%) Overall: cutting loss: 1.45%, field stacking: 0.50%, transportation loss from field to farmyard: 0.53%, threshing loss: 1.79%	Greeley (1982)

(continued)

Table A.7 (continued)

Method	Data year	Region(s)	Magnitude	Citation(s)
Questionnaire	–	Bangladesh	Mymensingh region: Harvesting loss: Aman season (2.45%), Boro season (2.47%); Aus season (3.00%) Threshing loss: Aman season (1.80%), Boro season (2.23%); Aus season (2.96%) Transportation loss: Aman season (1.59%), Boro season (2.01%); Aus season (1.64%) Khulna region: Harvesting loss: Aman season (1.54%), Boro season (1.40%); Aus season (0.21%) Threshing loss: Aman season (0.62%), Boro season (0.83%); Aus season (1.50%) Transportation loss: Aman season (0.69%), Boro season (1.03%); Aus season (0.25%) Dinajpur region: Harvesting loss: Aman season (1.51%), Boro season (1.78%); Aus season (2.06%) Threshing loss: Aman season (1.11%), Boro season (1.28%); Aus season (0.45%) Transportation loss: Aman season (0.62%), Boro season (0.93%); Aus season (0.84%) Comilla region: Harvesting loss: Aman season (1.07%), Boro season (1.02%); Aus season (0.89%) Threshing loss: Aman season (0.73%), Boro season (0.74%); Aus season (0.71%) Transportation loss: Aman season (0.63%), Boro season (0.71%); Aus season (0.54%) Nationwide: Harvesting loss: Aman season (1.60%), Boro season (1.62%); Aus season (1.91%) Threshing loss: Aman season (0.87%), Boro season (1.13%); Aus season (1.07%) Transportation loss: Aman season (1.10%), Boro season (1.22%); Aus season (1.79%)	Bala et al. (2010)

(continued)

Table A.7 (continued)

Method	Data year	Region(s)	Magnitude	Citation(s)
Field experiments	2013	Rangpur (Bangladesh)	Harvesting loss (shattering loss): Korean self-propelled reaper: 1.66%; China self-propelled reaper: 1.50%; BRRI reaper (power tiller-operated): 1.45%; manual reaping: 1.40%	Alam et al. (2018)
Field experiments	2018	Bangladesh	Manual operation (6.08%): shatter loss (0.74%), cutting loss (0.68%), gathering loss (0.31%), carrying loss (0.23%), threshing loss (3.35%), cleaning loss (0.78%) Combine harvester (from reaping to cleaning): 1.61%	Hasan et al. (2019)
Field experiments	2008–2010	Bangladesh	Aus season: BR26: Reaping loss by sickle (2.1%); Field transportation loss by trolley (0.2%), by head carry (0.65%), by shoulder carry (0.75%); threshing loss by ODT (1.21%), by CDT (1.98%); winnowing loss by Kula (0.25%), by winnower (0.25%) BRRI dhan27: Reaping loss by sickle (2.15%); Field transportation loss by trolley (0.195%), by head carry (0.84%), by shoulder carry (0.79%); threshing loss by ODT (1.1%), by CDT (1.2%); winnowing loss by Kula (0.17%), by winnower (0.26%) Aman season: BR23: Reaping loss by sickle (1.88%); Field transportation loss by trolley (0.16%), by head carry (0.49%), by shoulder carry (0.69%); threshing loss by ODT (1.07%), by CDT (2.27%); winnowing loss by Kula (0.22%), by winnower (0.24%) BR11: Reaping loss by sickle (2%); Field transportation loss by trolley (0.23%), by head carry (0.66%), by shoulder carry (0.63%); threshing loss by ODT (0.86%), by CDT (2.26%); winnowing loss by Kula (0.21%), by winnower (0.29%) Boro season: BRRI dhan28: Reaping loss by sickle (1.83%); Field transportation loss by trolley (0.15%), by head carry (0.51%), by shoulder carry (0.72%); threshing loss by ODT (1.1%), by CDT (2.14%); winnowing loss by Kula (0.17%), by winnower (0.28%) BRRI dhan29: Reaping loss by sickle (1.94%); Field transportation loss by trolley (0.24%), by head carry (0.81%), by shoulder carry (0.72%); threshing loss by ODT (1.13%), by CDT (1.96%); winnowing loss by Kula (0.18%), by winnower (0.16%) Cutting loss: Aus season (2.13%); Aman season (1.94%); Boro season (1.89%) Field stacking g loss: Aus season (0.69%); Aman season (0.97%); Boro season (0.83%) Field transportation loss: Aus season (0.57%); Aman season (0.48%); Boro season (0.53%) Threshing loss: Aus season (3.09%); Aman season (3.23%); Boro season (3.16%) Threshing loss: Aus season (0.47%); Aman season (0.48%); Boro season (0.39%)	Nath et al. (2016)

(continued)

Table A.7 (continued)

Method	Data year	Region(s)	Magnitude	Citation(s)
Questionnaire	2009–2010	Bangladesh	Reaping loss: Aman season (1.95 kg/quintal); Boro season (1.66 kg/quintal) Threshing loss: Aman season (0.64 kg/quintal); Boro season (0.56 kg/quintal) Winnowing loss: Aman season (0.32 kg/quintal); Boro season (0.24 kg/quintal)	Begum et al. (2012)
Literature review	2010	China	Combine harvesting: 1.5% Segmented harvesting: 4.4% Harvesting loss (average of combine harvesting and segmented harvesting): 2.7% Package transportation loss (from field to homestead or storage): 1% Bulk transportation loss (from field to homestead or storage): 0.3% Transportation loss (average of package transportation and bulk transportation): 0.9%	Gao et al. (2016)

(continued)

Table A.7 (continued)

Method	Data year	Region(s)	Magnitude	Citation(s)
Questionnaire	2014	China	Farmers' perceptions about harvest loss (from cutting, threshing, to packaging) in nationwide: 26.93% of farmers thought: less than 3% 29.20% of farmers thought: 3–4% 18.30% of farmers thought: 4–5% 13.07% of farmers thought: 5–6% 5.68% of farmers thought: 6–7% 6.82% of farmers thought: over 7% Farmers' perceptions about harvest loss (from cutting, threshing, to packaging) in east region: 26.38% of farmers thought: less than 3% 23.77% of farmers thought: 3–4% 18.55% of farmers thought: 4–5% 12.75% of farmers thought: 5–6% 6.38% of farmers thought: 6–7% 12.17% of farmers thought: over 7% Farmers' perceptions about harvest loss (from cutting, threshing, to packaging) in central region: 24.39% of farmers thought: less than 3% 32.52% of farmers thought: 3–4% 13.82% of farmers thought: 4–5% 16.67% of farmers thought: 5–6% 8.13% of farmers thought: 6–7% 4.47% of farmers thought: over 7% Farmers' perceptions about harvest loss (from cutting, threshing, to packaging) in west region: 19.28% of farmers thought: less than 3% 35.43% of farmers thought: 3–4% 26.46% of farmers thought: 4–5% 13.00% of farmers thought: 5–6% 3.59% of farmers thought: 6–7% 2.24% of farmers thought: over 7% Farmers' perceptions about harvest loss (from cutting, threshing, to packaging) in northeast: 65.15% of farmers thought: less than 3% 22.72% of farmers thought: 3–4% 6.06% of farmers thought: 4–5% 1.52% of farmers thought: 5–6% 1.52% of farmers thought: 6–7% 3.03% of farmers thought: over 7%	Wu et al. (2017)

(continued)

Table A.7 (continued)

Method	Data year	Region(s)	Magnitude	Citation(s)
Field experiments		Zhejiang (China)	Sickle reaping: scattered loss (0.09%), stacking loss (0.21%), uncut loss (0.11%) Combine harvester: scattered loss (0.89%), uncut loss (0.95%) Pedal thresher: unthreshed loss (0.5%), splash loss (0.16%), scattered loss (0.16%) Electronical thresher: unthreshed loss (0.67%), splash loss (0.26%), scattered loss (0.59%) Combine harvester: unthreshed loss (1.32%), splash loss (0.34%), scattered loss (included in scattered loss in cutting stage)	Li et al. (1991)
Field experiments, questionnaire	2019	Heilongjiang (China)	Harvest loss in field experiments: Loss by Kubota, 25% moisture content, 0.25 m stubble height, 11–12 km/h harvest speed: 545.87 ± 5.26 kg/h m^2 Loss by Kubota, 25% moisture content, 0.25 m stubble height, 8–9 km/h harvest speed: 436.43 ± 17.75 kg/h m^2 Loss by Kubota, 25% moisture content, 0.25 m stubble height, 5–6 km/h harvest speed: 171.46 ± 1.62 kg/h m^2 Loss by Kubota, 21% moisture content, 0.17 m stubble height, 11–12 km/h harvest speed: 108.7 ± 3.36 kg/h m^2 Loss by Kubota, 21% moisture content, 0.17 m stubble height, 8–9 km/h harvest speed: 46.8 ± 1.98 kg/h m^2 Loss by Kubota, 21% moisture content, 0.17 m stubble height, 5–6 km/h harvest speed: 13.4 ± 0.98 kg/h m^2 Loss by Yanmar, 25% moisture content, 0.25 m stubble height, 11–12 km/h harvest speed: 530.67 ± 7.12 kg/h m^2 Loss by Yanmar, 25% moisture content, 0.25 m stubble height, 8–9 km/h harvest speed: 447.07 ± 7.49 kg/h m^2 Loss by Yanmar, 25% moisture content, 0.25 m stubble height, 5–6 km/h harvest speed: 253.87 ± 2.73 kg/h m^2 Loss by Yanmar, 21% moisture content, 0.17 m stubble height, 11–12 km/h harvest speed: 211.37 ± 11.3 kg/h m^2 Loss by Yanmar, 21% moisture content, 0.17 m stubble height, 8–9 km/h harvest speed: 80.40 ± 17.75 kg/h m^2 Loss by Yanmar, 21% moisture content, 0.17 m stubble height, 5–6 km/h harvest speed: 46.43 ± 1.62 kg/h m^2 Harvest loss by farmers' perception: 3–5%	Gu et al. (2020)

(continued)

Table A.7 (continued)

Method	Data year	Region(s)	Magnitude	Citation(s)
Field experiments	2016	China	Harvest loss (including reaping loss, threshing loss, winnowing loss, and transportation loss from field to homestead) Combine harvester: The Northeast Plain (3.02%), Yangtze River Basin (3.17%), Southeast Coast (4.12%). Average nationwide (3.44%) Segmented harvesting: The Northeast Plain (1.41%), Yangtze River Basin (1.81%), Southeast Coast (1.76%). Average nationwide (1.66%) Harvest loss nationwide: 3.02%	Huang et al. (2018)
Questionnaire	2016	China	Segmented harvesting: reaping loss (2.48%), threshing loss (0.76%), winnowing loss (0.42%), transportation loss (0.22%) Combine harvesting: loss from reaping to winnowing (3.27%), transportation (0.12%) Average nationwide harvest loss: 3.65%	Qu et al. (2021a)
Questionnaire	2016	China	Harvest loss (from reaping to field transportation): Small-scale farmer: 4.59% Middle-scale farmer: 3.90% Large-scale farmer: 2.60%	Qu et al. (2021b)
FLA methodology	2015	Democratic Republic of Timor-Leste	Manual harvesting loss: 3.5% in harvesting stage; 3.5% of the initial quantity; USD 3140 Transportation loss from field to homestead: 1.5% in the transportation stage; 1.45% of the initial quantity; USD 1300 Thresher machine loss: 5% in the threshing stage; 4.75% of the initial quantity; USD 4260 Manual winnowing: 0.5% in the winnowing stage; 0.45% of the initial quantity; USD 400	FAO (2018)
Field experiments	2013	India	Reaping: 2.08% ± 0.79%, collection (including stacking, bundling and transportation up to threshing floor): 0.37% ± 0.29%, threshing: 1.44% ± 0.39%, winnowing/cleaning: 0.5% ± 0.5%	Jha et al. (2015)

(continued)

Table A.7 (continued)

Method	Data year	Region(s)	Magnitude	Citation(s)
Questionnaire	2010–2012	West Bengal (India)	Harvest loss: 17.45 kg/acre; 0.78 kg/quintal; 0.78% of harvest amount Mechanical threshing loss: 7.04 kg/acre; 0.31 kg/quintal; 0.31% of threshed amount Manual winnowing loss: 2.94 kg/acre; 0.13 kg/quintal; 0.13% of winnowed amount Transportation loss: head load (0.04 kg/quintal; 0.04% of amount transported), bullock cart (0.34 kg/quintal; 0.34% of amount transported), trolley (0.43 kg/quintal; 0.43% of amount transported), tempo (0.15 kg/quintal; 0.15% of amount transported) Harvest loss by farm size: Marginal size: 0.96 kg/quintal; small size: 0.85 kg/quintal; medium size: 0.74 kg/quintal; large size: 0.58 kg/quintal Average of four sizes: 0.78 kg/quintal Threshing loss by farm size: Marginal size: 0.46 kg/quintal; small size: 0.34 kg/quintal; medium size: 0.28 kg/quintal; large size: 0.23 kg/quintal Average of four sizes: 0.32 kg/quintal Winnowing loss by farm size: Marginal size: 0.20 kg/quintal; small size: 0.15 kg/quintal; medium size: 0.12 kg/quintal; large size: 0.10 kg/quintal Average of four sizes: 0.13 kg/quintal Transportation loss by farm size: Marginal size: 0.71 kg/quintal; small size: 0.61 kg/quintal; medium size: 0.52 kg/quintal; large size: 0.39 kg/quintal Average of four sizes: 0.55 kg/quintal	Sarkar et al. (2013)

(continued)

Table A.7 (continued)

Method	Data year	Region(s)	Magnitude	Citation(s)
Questionnaire	2011–2012	Assam (India)	Manual harvest loss (early season): Local paddy: 10.83 kg/hectare; 0.25 kg/quintal; 0.25% of harvest amount HYV paddy: 15.89 kg/hectare; 0.59 kg/quintal; 0.59% of harvest amount Manual harvest loss (mid-season): Local paddy: 12.89 kg/hectare; 0.40 kg/quintal; 0.40% of harvest amount HYV paddy: 25.57 kg/hectare; 0.56 kg/quintal; 0.56% of harvest amount Manual harvest loss (late-season): Local paddy: 16.42 kg/hectare; 0.98 kg/quintal; 0.98% of harvest amount HYV paddy: 54.69 kg/hectare; 0.96 kg/quintal; 0.96% of harvest amount Threshing loss (average of manual and mechanical): Local paddy: 0.45 kg/hectare; 1.04 kg/quintal; 1.04% of threshed amount HYV paddy: 0.99 kg/hectare; 1.50 kg/quintal; 1.50% of threshed amount Winnowing loss (average of manual and mechanical): Local paddy: 0.43 kg/hectare; 1.01 kg/quintal; 1.01% of threshed amount HYV paddy: 0.85 kg/hectare; 0.96 kg/quintal; 0.96% of threshed amount Transportation loss from field to homestead: head load (1.38 kg/quintal; 1.38% of amount transported), bullock cart (0.00 kg/quintal; 0.00% of amount transported), trolley (1.85 kg/quintal; 1.85% of amount transported), tempo (0.00 kg/quintal; 0.00% of amount transported), mini truck (1.90 kg/quintal; 1.90% of amount transported), hand cart (1.57 kg/quintal; 1.57% of amount transported). Total (1.67 kg/quintal; 1.67% of amount transported) Harvest loss by farm size: Marginal size: 0.48 kg/quintal; small size: 0.58 kg/quintal; medium size: 0.62 kg/quintal; large size: 0.81 kg/quintal Average of four sizes: 0.62 kg/quintal Threshing loss by farm size: Marginal size: 0.91 kg/quintal; small size: 0.98 kg/quintal; medium size: 1.41 kg/quintal; large size: 1.78 kg/quintal Average of four sizes: 1.27 kg/quintal Winnowing loss by farm size: Marginal size: 0.79 kg/quintal; small size: 0.88 kg/quintal; medium size: 1.02 kg/quintal; large size: 1.22 kg/quintal Average of four sizes: 0.98 kg/quintal Transportation loss by farm size: Marginal size: 1.30 kg/quintal; small size: 1.49 kg/quintal; medium size: 1.79 kg/quintal; large size: 2.11 kg/quintal Average of four sizes: 1.67 kg/quintal	Bordoloi (2013)

(continued)

Table A.7 (continued)

Method	Data year	Region(s)	Magnitude	Citation(s)
Questionnaire	2011–2012	Karnataka (India)	Harvest loss (including combine harvester and manual harvesting): Early season: 33.2 kg/acre; 1.74 kg/quintal; 1.74% of harvest amount Mid-season: 37.4 kg/acre; 1.92 kg/quintal; 1.92% of harvest amount Late-season: 41.4 kg/acre; 1.86 kg/quintal; 1.86% of harvest amount Total: 37.1 kg/acre; 1.90 kg/quintal; 1.90% of harvest amount Manual threshing loss: 21.92 kg/acre; 1.24 kg/quintal Mechanical threshing loss: 25.37 kg/acre; 1.16 kg/quintal Manual winnowing loss: 20.21 kg/acre; 1.14% of winnowed amount Mechanical winnowing loss: 8.33 kg/acre; 0.46% of winnowed amount Transportation loss (from field to homestead or market): head load (0.38 kg/quintal; 0.38% of amount transported), bullock cart (0.62 kg/quintal; 0.62% of amount transported), trolley (0.64 kg/quintal; 0.64% of amount transported), truck (0.80 kg/quintal; 0.80% of amount transported). Total (0.64 kg/quintal; 0.64% of amount transported) Harvest loss by farm size: Marginal size: 2.32 kg/quintal; small size: 1.80 kg/quintal; medium size: 1.99 kg/quintal; large size: 1.26 kg/quintal Average of four sizes:1.90 kg/quintal Threshing loss by farm size: Marginal size: 0.48 kg/quintal; small size: 0.17 kg/quintal; medium size: 0.11 kg/quintal; large size: 0.00 kg/quintal Average of four sizes: 0.20 kg/quintal Winnowing loss by farm size: Marginal size: 0.16 kg/quintal; small size: 0.12 kg/quintal; medium size: 0.04 kg/quintal; large size: 0.00 kg/quintal Average of four sizes: 0.08 kg/quintal Transportation loss by farm size: Marginal size: 0.84 kg/quintal; small size: 0.39 kg/quintal; medium size: 0.55 kg/quintal; large size: 0.52 kg/quintal Average of four sizes: 0.57 kg/quintal	Kannan et al. (2013)

(continued)

Table A.7 (continued)

Method	Data year	Region(s)	Magnitude	Citation(s)
Questionnaire	2011–2012	Punjab (India)	Mechanical harvest loss (HYV paddy): Early stage: 93.70 kg/acre; 3.40 kg/quintal; 3.40% of harvested amount Mid-season: 38.30 kg/acre; 1.40 kg/quintal; 1.40% of harvested amount Late-season: 53.60 kg/acre; 1.90 kg/quintal; 1.90% of harvested amount Transportation loss by tractor-trolley (to the market): 0.063 kg/quintal; 0.0002% of amount transported Harvest loss by farm size: Marginal size: 1.19 kg/quintal; small size: 1.66 kg/quintal; medium size: 1.64 kg/quintal; large size: 1.52 kg/quintal Average of four sizes:1.54 kg/quintal Transportation loss by farm size: Marginal size: 0.09 kg/quintal; small size: 0.09 kg/quintal; medium size: 0.05 kg/quintal; large size: 0.06 kg/quintal Average of four sizes: 0.06 kg/quintal	Grover et al. (2012)

(continued)

Table A.7 (continued)

Method	Data year	Region(s)	Magnitude	Citation(s)
Questionnaire	2010–2012	Tamil Nadu (India)	Tiruvarur district: Harvest loss of HYV paddy (including mechanical (over 90%) and manual harvesting): Early season: 92.32 kg/acre; 3.87 kg/quintal; 3.87% of harvest amount Mid-season: 51.2 kg/acre; 2.21 kg/quintal; 2.21% of harvest amount Late-season: 87.63 kg/acre; 3.68 kg/quintal; 3.68% of harvest amount Threshing loss of HYV paddy (including mechanical and manual threshing): 2.11 kg/quintal Winnowing loss of HYV paddy (including mechanical and manual threshing): 0.18 kg/quintal Transportation loss by tempo (to the market): 0.56 kg/quintal Villupuram district: Harvest loss of HYV paddy: Early season (mechanical harvesting): 71.75 kg/acre; 2.96 kg/quintal; 2.96% of harvest amount Mid-season (mostly mechanical harvesting): 69.79 kg/acre; 2.87 kg/quintal; 2.87% of harvest amount Late-season (mechanical harvesting): 89.3 kg/acre; 3.60 kg/quintal; 3.60% of harvest amount Threshing loss of HYV paddy (including mechanical and manual threshing): 0.83 kg/quintal Transportation loss by tempo (to the market): 0.65 kg/quintal Harvest loss by farm size: Tiruvarur district: marginal size: 3.12 kg/quintal; small size: 3.08 kg/quintal; medium size: 3.14 kg/quintal; large size: 3.07 kg/quintal; average of four sizes:3.10 kg/quintal Villupuram district: marginal size: 3.36 kg/quintal; small size: 3.19 kg/quintal; medium size: 2.94 kg/quintal; large size: 3.16 kg/quintal; average of four sizes:3.16 kg/quintal Average of two districts: 3.13% Threshing loss by farm size: Tiruvarur district: marginal size: 1.73 kg/quintal; small size: 1.57 kg/quintal; medium size: 2.77 kg/quintal; large size: 1.38 kg/quintal; average of four sizes: 2.11 kg/quintal Villupuram district: marginal size: 1.12 kg/quintal; small size: 1.07 kg/quintal; medium size: 0.46 kg/quintal; large size: 0.83 kg/quintal; average of four sizes:0.83 kg/quintal Average of two districts: 1.47% Winnowing loss by farm size: Tiruvarur district: marginal size: 0.15 kg/quintal; medium size: 0.46 kg/quintal; large size: 0.10 kg/quintal; average of four sizes:0.18 kg/quintal Transportation loss by farm size: Tiruvarur district: marginal size: 0.73 kg/quintal; small size: 0.56 kg/quintal; medium size: 0.50 kg/quintal; large size: 0.44 kg/quintal; average of four sizes:0.56 kg/quintal Villupuram district: marginal size: 0.84 kg/quintal; small size: 0.54 kg/quintal; medium size: 0.40 kg/quintal; large size: 0.65 kg/quintal; average of four sizes:0.65 kg/quintal Average of two districts: 0.61%	Sivagnanam (2013)

(continued)

Table A.7 (continued)

Method	Data year	Region(s)	Magnitude	Citation(s)
Questionnaire	2011–2012	Uttar Pradesh (India)	Harvest loss of HYV paddy (including mechanical and manual harvesting): Early season: 40.65 kg/acre; 0.92 kg/quintal; 0.92% of harvest amount Mid-season: 55.81 kg/acre; 3.69 kg/quintal; 3.69% of harvest amount Late-season: 36.36 kg/acre; 1.43 kg/quintal; 1.43% of harvest amount Winnowing loss of HYV paddy: 2.71 kg/acre, 1.28 kg/quintal, 1.28% of winnowed amount Transportation loss: head load (0.14 kg/quintal; 0.14% of amount transported), bullock cart (1.70 kg/quintal; 0.88% of amount transported), trolley (0.51 kg/quintal; 0.51% of amount transported), tempo (0.72 kg/quintal; 0.72% of amount transported). Total (0.49 kg/quintal; 0.49% of amount transported) Harvest loss by farm size: Marginal size: 2.53 kg/quintal; small size: 3.19 kg/quintal; medium size: 1.56 kg/quintal; large size: 2.45 kg/quintal Average of four sizes:2.71 kg/quintal Threshing loss by farm size: Marginal size: 1.78 kg/quintal; small size: 1.23 kg/quintal; medium size: 0.58 kg/quintal; large size: 0.98 kg/quintal Average of four sizes: 1.28 kg/quintal Winnowing loss by farm size: Marginal size: 0.64 kg/quintal; small size: 0.41 kg/quintal; medium size: 0.10 kg/quintal; large size: 0.16 kg/quintal Average of four sizes: 0.40 kg/quintal Transportation loss by farm size: Marginal size: 0.49 kg/quintal; small size: 0.62 kg/quintal; medium size: 0.41 kg/quintal; large size: 0.31 kg/quintal Average of four sizes: 0.48 kg/quintal	Roy (2013)
Questionnaire	2003–2004	Karnataka (India)	Harvesting loss: 0.40 kg/quintal, threshing loss: 0.52 kg/quintal, cleaning/winnowing loss: 0.20 kg/quintal, transportation loss: 0.50 kg/quintal	Basavaraja et al. (2007)
Field experiments	2010	Karnataka (India)	Combine harvester: 2.88%–3.60%	Veerangouda et al. (2010)
Field experiments	1981	Indonesia	Traditional ani-ani method: shattered and dropped losses (1.40%); uncut losses (4.48%); foot-treading threshing losses (2.38%) Sickle method: shattered and dropped losses (1.28%); uncut losses (1.92%); beating threshing losses (5.63%)	Gaiser and Esmay (1981)

(continued)

Table A.7 (continued)

Method	Data year	Region(s)	Magnitude	Citation(s)
Field experiments	2010	Iran	Quantitative loss: Reaping loss: T1 (1.60%); T2 (1.48%); T3 (1.54%) Threshing loss: T1 (0.98%); T2 (1.04%); T3 (1.12%) Reaping + threshing loss: T1 (2.58%); T2 (2.52%); T3 (2.26%); T4 (2.27%); T5 (2.4%) Qualitative loss: T1 (2.05%); T2 (2.44%); T3 (2.41%); T4 (0.47%); T5 (0.75%) The average quantitative loss of T1–T3: 2.58%; The average quantitative loss of T4–T5: 2.33% The average qualitative loss of T1–T3: 2.30%; The average qualitative loss of T4–T5: 0.61% Quantitative losses are the result of shattering and losing of grain and non-threshed panicles during reaping and threshing Qualitative losses are owing to broken, husked, and cracked grains from environmental and mechanical impacts T1–T3 are regarded as indirect harvesting, T4–T5 are regarded as direct harvesting Note: T1: Manual harvesting (cutting with sickle) + tractor-driven thresher T2: Rice reaper + tractor-driven thresher T3: Rice reaper + threshing by a universal combine equipped with pick-up header T4: Head-feed rice combine harvester T5: Whole-crop rice combine harvester	Alizadeh and Allameh (2013)
Field experiments	2014–2015	Myanmar	Wet season 2014: Harvesting loss: IPR (16.0%); FP1W (28.2%); FP4W (23.63%) Manual cutting and handing loss: IPR (13.6%); FP1W (20.8%); FP4W (14.4%) In-field stacking loss: FP1W (0.3%); FP4W (0.6%) Threshing loss: IPR (2.4%); FP1W (7.2%); FP4W (8.7%) Dry season 2015: Harvesting loss: IPRc (1.7%); FP (9.3%) Manual cutting and handing loss: FP (6.7%) Threshing loss: FP (2.6%) Combine harvesting loss: IPRc (1.7%) Dry season 2016: Harvesting loss: IPRc (0.9%); FP (4.0%) Manual cutting and handing loss: FP (1.8%) Threshing loss: FP (2.2%) Combine harvesting loss: IPRc (0.9%) Note: IPR: manual cutting, threshing immediately after cutting using improved thresher FP1W: manual cutting, stacking 1 week in field, less developed thresher FP4W: manual cutting, stacking 4 weeks in field, less developed thresher IPRc: combine harvester FP: manual cutting, threshing immediately after cutting using less improved thresher	Gummert et al. (2020)

(continued)

Table A.7 (continued)

Method	Data year	Region(s)	Magnitude	Citation(s)
–	–	Thailand	Average loss as a percentage of estimated potential yield: Traditional hand cutting loss: Thailand (9.3%) Shoulder power reaper: Thailand (5.2%) Reaper-binder: Thailand (5.2%) Combine harvester: Thailand (1.1%)	Grolleaud (2002)
Field experiments	1979	Dominican Republic	Harvest loss by region: Central-Northeast: 17.41% Northwest: 21.58% Southwest: 14.25% Harvest by size (tarea): 1–50: 18.24% 51–100: 24.82% 101 + : 12.27% Harvest loss by harvest method: Manual: 20.32% Mechanized: 13.37% Harvest loss by hand-threshing method: Stick: 19.52% Platform: 22.01% Drum: 17.72%	Boxall et al. (1981)

Note Countries or regions are sorted alphabetically
Sources See column 5 in table

References

Alizadeh MR, Allameh A (2013) Evaluating rice losses in various harvesting practices. Int Res J Apllied Basic Sci 4:894–901

Amusat MA, Eneh CK, Obiakor SC (2016) Assessment of postharvest losses of rice at different stages of operation. Int J Life Sci 5:50–53

Amponsah SK, Addo A, Dzisi K, et al (2018) Assessment of rice farmers' knowledge and perception of harvest and postharvest losses in Ghana. Cogent Food Agric 4:1471782. https://doi.org/10.1080/23311932.2018.1471782

Appiah F, Guisse R, Dartey PKA (2011) Post harvest losses of rice from harvesting to milling in Ghana. J Stored Prod Postharvest Res 2:64–71

Badawi AT (2001) A proposal on the assessment of rice post-harvest losses. In: The new development in rice agronomy and its effects on yield and quality in mediterranean areas. CIHEAM, Montpellier

Bala BK, Haque M a, Hossain MA, Majumdar S (2010) Post harvest loss and technical efficiency of rice, wheat and maize production system: assessment and measures for strengthening food security. Bangladesh Agricultural University, Bengaluru, India

Basavaraja H, Mahajanashetti SB, Udagatti NC (2007) Economic analysis of post-harvest losses in food grains in India: a case study of Karnataka. Agric Econ Res Rev 20:117–126. https://doi.org/10.22004/ag.econ.47429

Begum EA, Hossain MI, Papanagiotou E (2012) Economic analysis of post-harvest losses in food grains for strengthening food security in northern regions of Bangladesh. Int J Appl Res Bus Adm Econ 01:56–65

Bordoloi J (2013) Assessment of pre and post harvest losses of paddy and wheat in Assam. Agro-Economic Research Centre for North-East India Assam Agricultural University, Jorhat, India

Boxall RA, La Gra J, Martinez E, Martinez J (1981) Post harvest losses of rice in the Dominican Republic. Trop Stored Prod Inf 42:5–10

Candia A, Okurut S, Komaketch A, et al (2012) On-farm post-harvest physical grain losses of "Kaiso" rice variety in Eastern Uganda. Uganda J Agric Sci 13:61–70

Coker AA, Ninalowo SO (2016) Effect of post-harvest losses on rice farmers' income in Sub-saharan Africa: a case of Niger state, Nigeria. J Agric Sci Food Technol 2:27–34

Danbaba N, Idakwo PY, Kassum AL, et al (2019) Rice postharvest technology in Nigeria: An overview of vurrent status, constraints and potentials for sustainable development. Open Access Libr J 6:1–23. https://doi.org/10.4236/oalib.1105509

FAO (2018) Food loss analysis: causes and solutions—case study on the rice value chain in the Democratic Republic of Timor-Leste. Food and Agriculture Organization of the United Nations, Rome, Italy

Gaiser D, Esmay M (1981) Traditional rice harvest loss and labor values in Indonesia. Trans ASAE 24:1162–1166. https://doi.org/10.13031/2013.34413

Gao LW, Xu SW, Li ZM et al (2016) Main grain crops postharvest losses and its reducing potential in China. Trans Chinese Soc Agric Eng 32:1–11

Gu YN, Sun HY, Bi HW et al (2020) Effect of mechanical harvest on rice loss after mature in Heilongjiang. Agric Outlook 16:114–118

Guisse R (2010) Post harvest losses of rice (oriza spp) from harvesting to milling: A case study in Besease and Nobewam in the Ejisu Juabeng district in the Ashanti region of Ghana. Kwame Nkrumah University, Kabwe, Zambia

Gummert M, Nguyen-Van-Hung, Cabardo C et al (2020) Assessment of post-harvest losses and carbon footprint in intensive lowland rice production in Myanmar. Sci Rep 10:1–13. https://doi.org/10.1038/s41598-020-76639-5

Greeley M (1982) Farm-level post-harvest food losses: the myth of the soft third option. IDS Bull 13:51–60. https://doi.org/10.1111/j.1759-5436.1982.mp13003007.x

Grolleaud M (2002) Post-harvest losses: discovering the full story overview of the phenomenon of losses during the post-harvest system. Food and Agriculture Organization of United Nation, Rome, Italy

Grover DK, Singh JM, Singh P (2012) Assessment of pre and post harvest losses in wheat and paddy crops in Punjab. Agro-Economic Research Centre Department of Economics and Sociology Punjab Agricultural University, Ludhiana, India

Hasan MK, Ali MR, Saha CK, et al (2019) Combine harvester: Impact on paddy production in Bangladesh. J Bangladesh Agric Univ 17:583–591. https://doi.org/10.3329/jbau.v17i4.44629

Huang D, Yao L, Wu LP, Zhu X Di (2018) Measuring rice loss during harvest in China: Based on experiment and survey in five provinces. J Nat Resour 33:1427–1438. https://doi.org/10.31497/zrzyxb.20170810

Jha SN, Vishwakarma RK, Ahmad T et al (2015) Assessment of quantitative harvest and post-harvest losses of major crops/commodities in India. Ministry of Food Processing Industries, Ludhiana, India

Kannan E, Kumar P, Vishnu K, Abraham H (2013) Assessment of pre and post harvest losses of rice and red gram in Karnataka. Agricultural Development and Rural Transformation Centre, Bangalore, India

Li ZF, Xia PK, Wang ZH et al (1991) Analysis of the constitution of grain postproduction losses and the preventive measures. J Zhejiang Univ 17:389–395

Nath B, Hossen M, Islam A et al (2016) Postharvest loss assessment of rice at selected areas of Gazipur district. Bangladesh Rice J 20:23–32. https://doi.org/10.3329/brj.v20i1.30626

Ndindeng SA, Candia A, Mapiemfu-Lamare D et al (2021) Valuation of rice postharvest losses in Sub-Saharan Africa and its mitigation strategies. Rice Sci 28:212–216. https://doi.org/10.1016/j.rsci.2021.04.001

Oguntade AE, Thylmann D, Deimling S (2014) Post-harvest losses of rice in Nigeria and their ecological footprint. Federal Ministry of Economic Cooperation and Development, Deutsche Gesellschaft fur Internatinale Zusammenarbeit (GIZ), Bonn, Germany

Qu X, Kojima D, Nishihara Y et al (2021a) Can harvest outsourcing services reduce field harvest losses of rice in China? J Integr Agric 20:1396–1406. https://doi.org/10.1016/s2095-3119(20)63263-4

Qu X, Kojima D, Nishihara Y et al (2021b) A study of rice harvest losses in China : do mechanization and farming scale matter? Japanese J Agric Econ 23:83–88. https://doi.org/10.18480/jjae.23.0_83

Roy R (2013) Assessment of pre and post harvest losses in wheat and paddy crops in Uttar Pradesh. Agro-Economic Research Centre, University of Allahabad, Allahabad

Sarkar D, Datta V, Chattopadhyay KS (2013) Assessment of pre and post harvest losses in rice and wheat in West Bengal. Agro-Economic Research Centre, Visva-Bharati, Santiniketan

Sanneh L (2015) Effects of threshing and post-threshing recovery methods on postharvest losses in two varieties of rice. Kwame Nkrumah University, Kabwe, Zambia

Sivagnanam KJ (2013) Estimation of pre-and post-harvest losses in paddy crop in Tamil Nadu. Agro-Economic Research Centre, University of Madras, Chennai, India

Totobesola M, Delve R, Nkundimana J d'Amour et al (2022) A holistic approach to food loss reduction in Africa: food loss analysis, integrated capacity development and policy implications. Food Secur 14:1401–1415. https://doi.org/10.1007/s12571-021-01243-y

Veerangouda M, Sushilendra S, Prakash K V., Anantachar M (2010) Performance evaluation of tractor operated combine harvester. Karnataka J Agric Sci 23:282–285

Wu LH, Hu QP, Wang JH, Zhu D (2017) Empirical analysis of the main factors influencing rice harvest losses based on sampling survey data of ten provinces in China. China Agric Econ Rev 9:287–302. https://doi.org/10.1108/CAER-03-2016-0036

Index

X. Qu et al., *Harvest Loss in China*, The University of Tokyo Studies on Asia,
https://doi.org/10.1007/978-981-97-9156-9

SPRINGER NATURE

GPSR Compliance

The European Union's (EU) General Product Safety Regulation (GPSR) is a set of rules that requires consumer products to be safe and our obligations to ensure this.

If you have any concerns about our products, you can contact us on ProductSafety@springernature.com

In case Publisher is established outside the EU, the EU authorized representative is:

Springer Nature Customer Service Center GmbH
Europaplatz 3
69115 Heidelberg, Germany

The manufacturer's authorised representative in the EU is Springer Nature Customer Service Centre GmbH, Europaplatz 3, 69115 Heidelberg, Germany. If you have any concerns regarding our products, please contact ProductSafety@springernature.com

Printed and bound by CPI Group (UK) Ltd, Croydon, CR0 4YY
01/10/2026
02225914-0001